LIPOXINS

Biosynthesis, Chemistry, and
Biological Activities

ADVANCES IN EXPERIMENTAL MEDICINE AND BIOLOGY

Recent Volumes in this Series

LIPOXINS

Biosynthesis, Chemistry, and
Biological Activities

Edited by

Patrick Y-K Wong

New York Medical College
Valhalla, New York

and

Charles N. Serhan

Brigham and Women's Hospital
and Harvard Medical School
Boston, Massachusetts

SPRINGER SCIENCE+BUSINESS MEDIA, LLC

Library of Congress Cataloging in Publication Data

Lipoxins: biosynthesis, chemistry, and biological activities.

(Advances in experimental medicine and biology; v. 229)
Proceedings of a FASEB symposium entitled "Lipoxins: biosynthesis and pharma-
cology," held March 29–April 3, 1987, in Washington, D.C.
Includes bibliographies and index.
1. Lipoxins — Congresses. I. Wong, Patrick, Y-K   II. Serhan, Charles N. III. Federa-
tion of American Societies for Experimental Biology. IV. Series. [DNLM: 1. Lipoxy-
genases — congresses. W1 AD559 v.229 / QU 140 L764 1987]
QP752.L52L56   1988   574.19′247   88-2544
ISBN 978-1-4757-0939-1   ISBN 978-1-4757-0937-7 (eBook)
DOI 10.1007/978-1-4757-0937-7

Proceedings of a FASEB symposium on Lipoxins: Biosynthesis and Pharmacology,
held March 29–April 3, 1987, in Washington, D.C.

© 1988 Springer Science+Business Media New York
Originally published by Plenum Press, New York in 1988

PREFACE

The discovery of new structures which display biological activities is always an exciting event in biomedical research. In recent years, advances in this area have occurred at a rapid pace. This is particularly evident in the field of eicosanoid research because of the close interactions between chemists and biomedical researchers. The lipoxins are a new class of lipid-derived oxygenation products, discovered in 1984, which can originate from either arachidonic acid or eicosapentaenoic acid. It is now clear that these compounds can be generated by sequential lipoxygenation of either arachidonic acid or eicosapentaenoic acid within various cells or during cell-cell interactions. Continued research on the total synthesis of these and related compounds, their biosynthesis, biological roles and mechanism(s) of action may contribute to the development of new therapeutic agents.

This volume contains chapters from lectures given at the first symposiums devoted to this area held at the 1987 FASEB Meeting in Washington, D.C. entitled "Lipoxins: Biosynthesis and Pharmacology". In addition to chapters from these presentations, several other chapters are included by other investigators who have contributed to this rapidly growing area. It is our intention that this volume represents a complete and up-to-date collection of the most recent information regarding the Lipoxins.

The Editors

ACKNOWLEDGEMENTS

We wish to express our gratitude to the American Society of Pharmacology and Experimental Therapeutics for their advice and assistance during the organization of this Symposium. And also, we wish to acknowledge with thanks the generous financial support from the following pharmaceutical companies in the U.S.A. They are Ciba Geigy Inc., The Upjohn Company, S.K.F. and Beckman Inc., Ortho Pharmaceutical Corp., Lederle Inc. and W.W. Diagnostic Inc.

The organizers of this Symposium hope that the chapters of this volume will serve as a guided reference to stimulate further studies and new developments in this area.

The Editors

CONTENTS

PART III. ACTIONS OF LIPOXINS OF THE 4- AND 5- SERIES

LIPOXINS: A NEW SERIES OF EICOSANOIDS

(BIOSYNTHESIS, STEREOCHEMISTRY, AND BIOLOGICAL ACTIVITIES)

Charles N. Serhan[1] and Bengt Samuelsson[2]

[1]Hematology Division
 Brigham and Women's Hospital and
 Harvard Medical School
 Boston, Mass.

[2]Department of Physiological Chemistry
 Karolinska Institutet
 Stockholm, Sweden

ABSTRACT

The oxygenation of arachidonic acid and other polyunsaturated fatty acids by a wide variety of cell types results in the formation of several structurally distinct classes of biologically active compounds[1,2]. These compounds include the prostaglandins, thromboxanes, leukotrienes, and other oxygenated derivatives of polyunsaturated fatty acids. A most recent addition to this family of biologically active compounds is the lipoxins (Figure 1). Leukotrienes and lipoxins are formed by mechanisms which involve initial oxygenation of free fatty acids by lipoxygenases. In general, lipoxygenase products display a wide range of actions and appear to be involved in immunity, the regulation of inflammation, and other physiological and pathophysiological processes. In this chapter we describe results of recent studies on the isolation, biosynthesis, stereochemistry and biological activities of this new series of compounds (lipoxins).

Isolation of the Lipoxins

Since lipoxygenation of arachidonic acid results in the formation of products of importance both in normal and pathological events (reviewed in refs. 1,2) bioregulation and interactions along these enzymatic pathways are of considerable interest. Whereas the biosynthesis of leukotrienes is initiated at the C-5 position of arachidonic acid, results from many studies suggested that initial lipoxygenation at the C-15 position can lead to the formation of compounds that may be of biological interest[2]. In particular, 15(S)-hydroxy-5,8,11-cis-13-trans-eicosatetraenoic acid (15-HETE) has been identified as a major product of arachidonic acid metabolism in both normal and asthmatic human lung tissue[3]. This lipoxygenase product has also been observed in large amounts in bronchoalveolar lavage fluids from patients with chronic stable asthma following antigenic challenge[4]. Taken together these findings suggest that products derived from the action of a 15-lipoxygenase on arachidonic acid may play a role in human pathophysiology.

Although the 15-lipoxygenase activity is a major route of arachidonic acid metabolism in a wide variety of mammalian tissues (reviewed in ref. 5), the receptor-mediated activation of this enzyme system and general physiological role of its products remains a subject of considerable interest. Our initial studies on the metabolism of [1-14C]-arachidonate in suspensions of mixed human leukocytes (i.e., neutrophils, eosinophils, basophils, etc.) indicated that a large percentage of the label material was transformed and associated with polar compounds which had not been previously described. To mimic cellular events and the reaction sequence(s) which may have given rise to these polar compounds, as well as to study interactions between the lipoxygenase pathways, we prepared both 15(S)-hydroperoxy-5,8,11-cis-13-trans-eicosatetraenoic acid (15-HPETE) and 15-HETE and studied the products formed upon incubation of these materials with human leukocytes. These experiments led to the isolation of a new series of biologically active oxygenated derivatives of arachidonic acid which contain a conjugated tetraene structure as a characteristic feature of the group. Since these compounds arose via interaction(s) between lipoxygenase pathways, we proposed the name lipoxins (lipoxygenase interaction products) for this group[6,7].

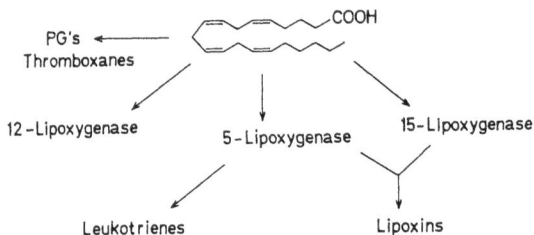

Fig. 1. Transformations of arachidonic acid.

Formation of these compounds was increased when the ionophore A23187 was added along with 15-HPETE incubations of human leukocytes. Following purification by silicic acid chromatography and thin layer chromatography, a fraction containing several unidentified tetraene-containing compounds was obtained from leukocyte suspensions. Samples of these materials were esterified, separated by thin layer chromatography, and analyzed by reversed-phase high pressure liquid chromatography. The basic structures of two main compounds of this series were elucidated by physical methods which included ultraviolet spectrometry, gas chromatography - mass spectrometry (utilizing several derivatives), and oxidative ozonolysis. One compound was identified as 5,6,15 L-trihydroxy-7,9,11,13-eicosatetraenoic acid, and the other as 5D,14,15-trihydroxy-6,8,10,12-eicosatetraenoic acid[2,6-9]. Addition of these biologically derived materials to either human neutrophils or human natural killer (NK) cells provoked selective responses different than those obtained with either leukotrienes or other eicosanoids. Hence the compounds were termed lipoxin A_4 (LXA$_4$) and lipoxin B_4 (LXB$_4$) respectively[7-9].

Stereochemistry and Biosynthesis

Next, it was of importance to determine both the complete stereochemistry of these compounds and their naturally occurring isomers as well
as explore their route(s) of biosynthesis. To this end, human leukocytes in
the presence and absence of ionophore were exposed to either 15-HPETE or
15-HETE and the trihydroxytetraene compounds were isolated and characterized
(a schematic summary is presented in figures 2 and 3). Here, exposure to
15-HPETE alone led to both activation of a 5-lipoxygenase activity and to
consumption of 15-HPETE to form tetraene-containing compounds. Addition of
15-HETE to cells exposed to either the ionophore A23187 or the chemotactic
peptide f-met-leu-phe led to the formation of these compounds and reduced
the appearance of products formed by non-enzymatic degradation of 15-HPETE
which can be formed under similar conditions[7,9]. Human leukocytes incubated
with 15-HETE in the absence of various stimuli did not generate lipoxins
suggesting that activation of these cells is required for the utilization
and subsequent transformation of 15-HETE to form lipoxins[9-12].

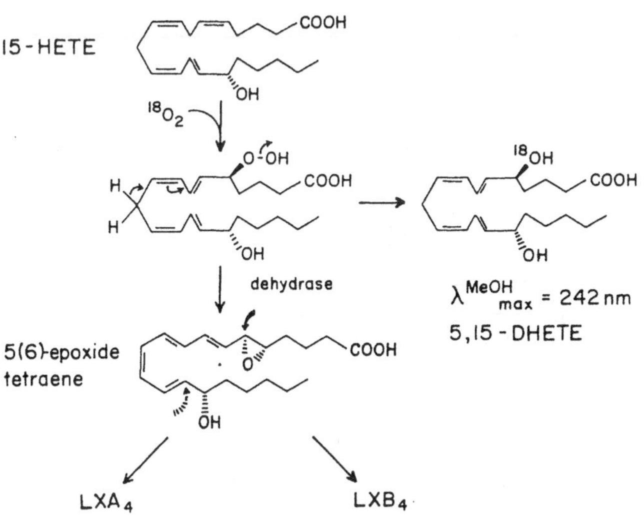

Fig. 2. Scheme of formation of lipoxins. When activated, human
leukocytes convert 15-HETE by a 5-lipoxygenase activity to
5(S)-hydroperoxy-15(S)-hydroxy-6,13-trans-8,11-cis-eicosa-
tetraenoic acid which can be further transformed to a
5(6)-epoxidetetraene intermediate[11,12]. In the presence of
isotopic oxygen each of the compounds carried an ^{18}O atom at
the carbon 5 position[10]. The 5(6)-epoxide tetraene (one
proposed intermediate is 15(S)-hydroxy-5,6-epoxy-7,9,13-trans-
11-cis-eicosatetraenoic acid) can be enzymatically converted
to lipoxin A4 by the action of an epoxide hydrolase (black
arrow) or by attack of the carbon-14 position (hatched arrow)
with the generation of an 8-cis double bond to form lipoxin
B4.

Strict criteria and synthetic materials prepared by total synthesis were employed to establish the complete stereochemistry of these and related compounds. The synthetic and biologically-derived materials were both subject to analysis by ultraviolet spectroscopy, HPLC (isochromatography in several systems), gas chromatography-mass spectroscopy of several derivatives, and bioassay[11-13]. Comparisons with several synthetic 5,14,15-trihydroxyeicosatetraenes prepared by Dr. J. Morris (The Upjohn Company, Kalamazoo, Mich.) showed that biologically derived LXB$_4$ is 5S,14R,15S-trihydroxy-6,10,12-trans-8-cis-eicosatetraenoic acid. The two naturally occurring isomers of LXB$_4$ were shown to be 5S,14R,15S-trihydroxy-6,8,10,12-trans-eicosatetraenoic acid (8-trans-LXB$_4$) and 5S,14S,15S-trihydroxy-6,8,10,12-trans-eicosatetraenoic acid (14S-8-trans-LXB$_4$)[10,11].

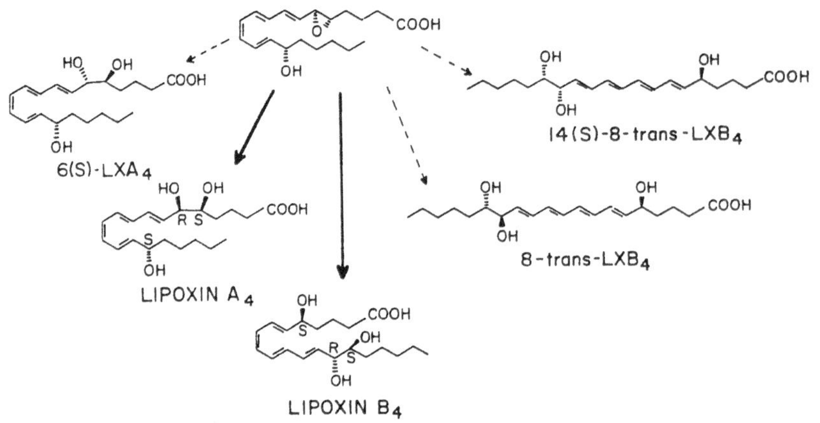

Fig. 3. One biosynthetic pathway for lipoxin formation via a 15(S)-hydroxy-5(6)-oxido-7,9,13-trans-11-cis-eicosatetraenoic acid. The stereochemistry of all of the compounds shown has been determined[11,12].

A synthetic approach was also undertaken to establish the complete stereochemistry of the biologically derived lipoxin A$_4$[12]. In collaboration with Prof. K.C. Nicolaou and his colleagues Dr. S.E. Webber and Dr. C.A. Veale of the University of Pennsylvania, studies with several synthetic 5,6,15-trihydroxyeicosatetraenoic acids demonstrated that the biologically derived LXA$_4$ is 5S,6R,15S-trihydroxy-7,9,13-trans-11-cis-eicosatetraenoic acid. The 6S isomer of LXA$_4$ (6S-LXA$_4$) was also identified from leukocyte extracts as well as two all-trans isomers which were designated 6S-11-trans-LXA$_4$ and 11-trans-LXA$_4$. The results of further studies indicated that these all-trans isomers arise, at least in part, by isomerization upon isolation and workup of LXA$_4$ and its epimer[10,12].

To shed light on possible routes of biosynthesis we examined the origins of oxygen in these compounds[10]. Here, incubations were performed under an atmosphere enriched in isotopic oxygen with activated human leukocytes exposed to either 15-HPETE or 15-HETE. LXA$_4$, LXB$_4$ and 5,15-DHETE, as well as the other tetraene-containing isomers, were isolated and analyzed (Figs. 2-5). Results from these studies demonstrated the incorporation of ^{18}O into each of the compounds and established that they each carried an ^{18}O atom at the carbon-5 position. In addition, they showed

4

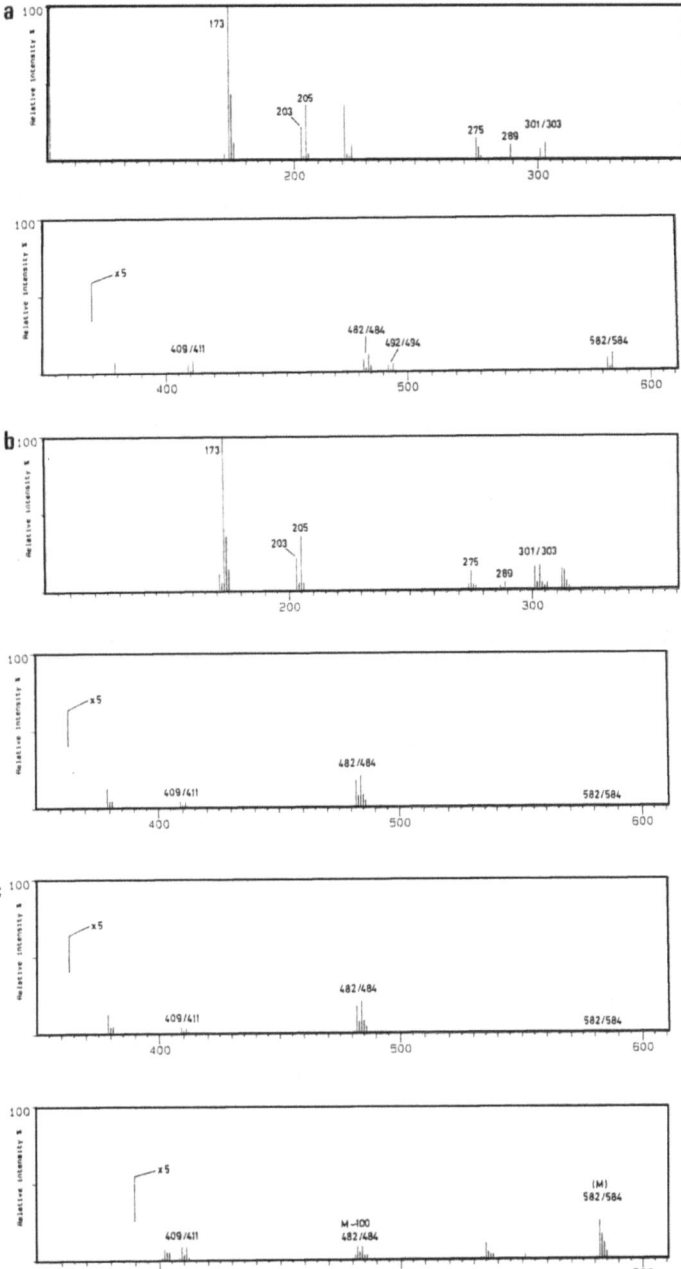

Fig. 4. Mass spectra of the Me3Si derivatives of the methyl esters of
LXB$_4$, 14S-8-trans-LXB$_4$, and 8-trans-LXB$_4$ obtained from human
leukocytes exposed to 15-HETE (50 µM) and A23187 (2.5 µM)
incubated in an atmosphere rich in $^{18}O_2$. ^{18}O-labeled Me3Si
derivative of the methyl ester of A) 14S-8-trans-LXB$_4$,
B) 8-trans-LXB$_4$, C) LXB$_4$. The prominent ions and their
positions are indicated. The ion m/e 203, which contains
carbons originating from C-1 through C-5 positions, were in
each case shifted to m/e 205 indicating the incorporation of
^{18}O at carbon 5. In addition, the oxygen atoms at carbon 14
positions were not derived from $^{18}O_2$, since the prominent ions
showed shifts of M+2 rather than M+4.

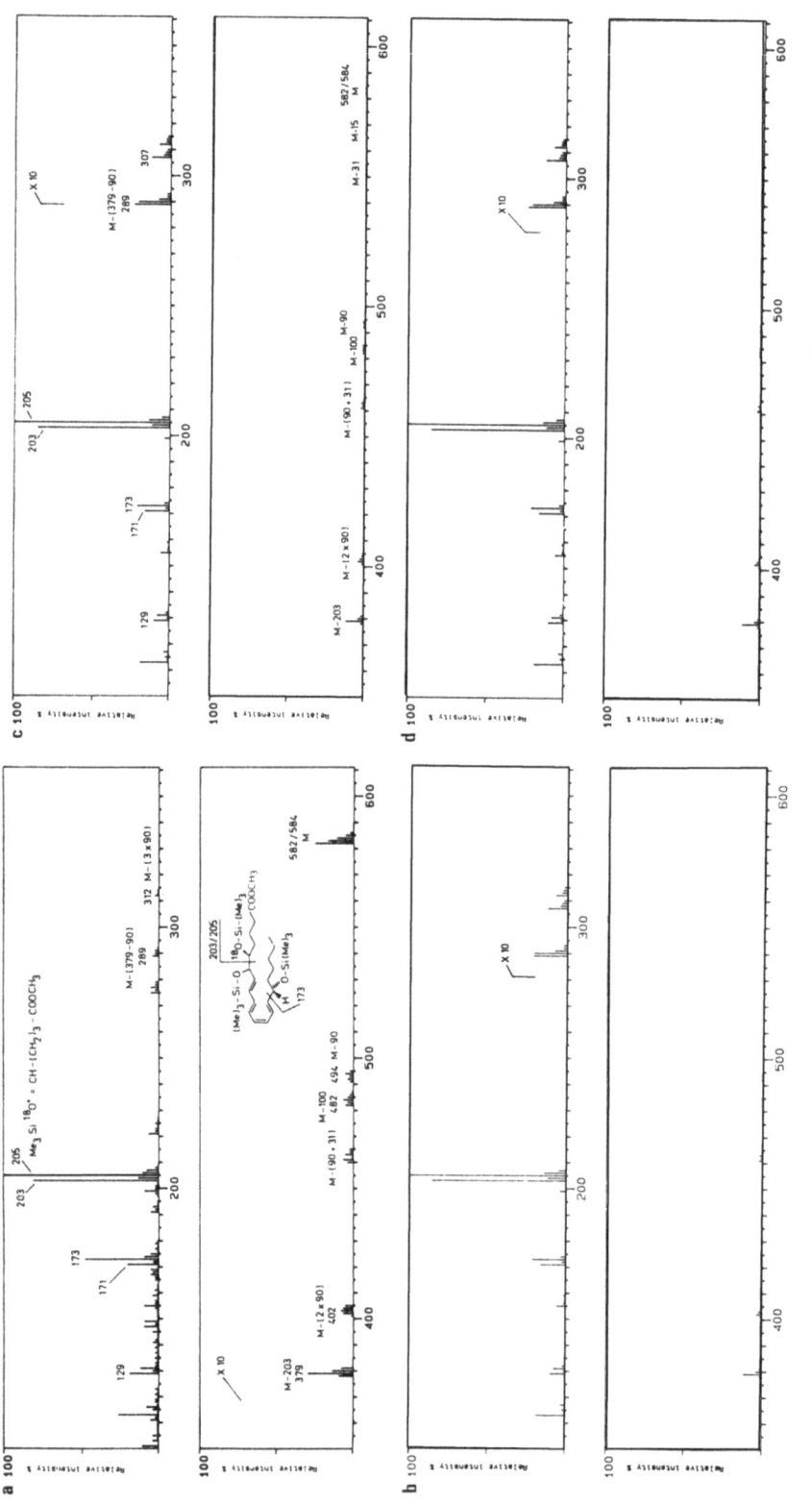

Fig. 5. Mass spectra of the Me₃Si derivatives of LXA₄ and its isomers obtained from human leukocyctes maintained under an atmosphere rich in $^{18}O_2$ and incubated with 15-HETE (50 µM) and A23187 (2.5 µM). ^{18}O-labeled Me₃Si derivative of the methyl ester of A) LXA₄, B) 11-trans LXA₄, C) 6s-11-trans-LXA₄ and D) 6s-LXA₄. The prominent ions and their positions are indicated. Again, ^{18}O was incorporated into the C-5 position of LXA₄ and each isomer 10,12.

that the oxygen atoms at either carbon-6 of LXA_4 or carbon-14 of LXB_4 as well as their isomers were not exclusively derived from $^{18}O_2$. Results obtained with either 15-HPETE or 15-HETE were virtually identical. These findings and results of alcohol trapping studies suggested the involvement of unstable epoxide intermediates in the formation of both LXA_4 and LXB_4[11-14].

Although it is clear that several distinct biosynthetic routes can be involved in the formation of tetraene containing eicosanoids[6,7], the finding that 15-HETE is also transformed by activated leukocytes to these products provided us with a model for studying a more limited biosynthetic path operative in their formation. Moreover, this finding provides a basis for exploring cell-cell interactions in the formation of lipoxin A_4 and lipoxin B_4 (e.g. transcellular metabolism of 15-HETE)[11,12]. In this route, schematically summarized in Figure 2, 15-HETE is converted to 5(S)-hydroperoxy-15(S)-hydroxy-6,13-trans-8,11-cis-eicosatetraenoic acid by activated cells which is further transformed to a 5(6)-epoxide tetraene. One proposed intermediate is 15(S)-hydroxy-5,6-epoxy-7,9,13-trans-11-cis-eicosatetraenoic acid[10-12]. Such an epoxide or its equivalent could be enzymatically transformed to either lipoxin A_4 (by the action of an epoxide hydrolase) or lipoxin B_4 (by attack of the C-14 position with the generation of an 8-cis double bond) (Figs. 2 and 3). Other isomers may be generated by non-enzymatic hydrolysis of the 5(6)-epoxytetraene or by isomerizations of LXA_4 or LXB_4 from conditions encountered upon isolation or by interactions of these compounds with metal-containing proteins[11,12].

This scheme of events is supported by several lines of evidence:

(1) 15-HETE serves as a precursor for formation of both LXA_4 and LXB_4 in activated leukocytes (which excludes the involvement of an epoxide intermediate at the 14(15) position);
(2) the pattern of isotopic oxygen incorporation in 5,15-DHETE, LXA_4, LXB_4, and their isomers (Figs. 4 and 5);
(3) the absolute stereochemistry of LXA_4 and LXB_4;
(4) time course of formation of these compounds by leukocytes (Fig. 6); and
(5) identification of alcohol trapping products (i.e. 15-HETE derived 5,15-dihydroxy-14-O-alkyleicosatetraenoic acids) originating from a 5(6)-epoxide tetraene (13,14).

Further evidence for the role of a 5(6)-epoxytetraene intermediate in the biosynthesis of these compounds was obtained by preparing 15(S)-hydroxy-5(6)-oxido-7,9,13-trans-11-cis-eicosatetraenoic acid by total chemical synthesis[13,14]. When added to purified human liver cytosolic epoxide hydrolase, the synthetic epoxide was rapidly (less than 5 seconds) and quantitatively converted into lipoxin A_4[14]. This system provides a clear model for evaluating the enzymatic formation of lipoxin A_4 (Fig. 7). It remains to be determined whether a similar enzyme is solely responsible for the formation of LXA_4 by human leukocytes. Others have also postulated the role of epoxide tetraenes in the formation of lipoxins and related compounds[15-20] and have isolated lipoxins of the 5 series which are formed from eicosapentaenoic acid[21]. In accordance with this observation we recently proposed that tetraene-containing compounds derived from arachidonic acid be denoted as lipoxins (LX) of the four series (i.e. lipoxin A_4 or LXA_4 and lipoxin B_4 or LXB_4) and those derived from eicosapentaenoic acid be termed lipoxins of the five series (i.e. lipoxin A_5 or LXA_5 and lipoxin B_5 or LXB_5) rather than lipoxenes[22].

In addition to the above-mentioned route of lipoxin formation, we have recently found that eosinophil rich granulocyte suspensions obtained from the peripheral blood of eosinophilic donors can generate LXA_4 from endogenous sources of arachidonate when exposed (in vitro) to ionophore A23187 (Fig. 8). In this study, neither lipoxin B_4 nor 6S-LXA_4 were consistently detected in extracts from these incubations. These eosinophil

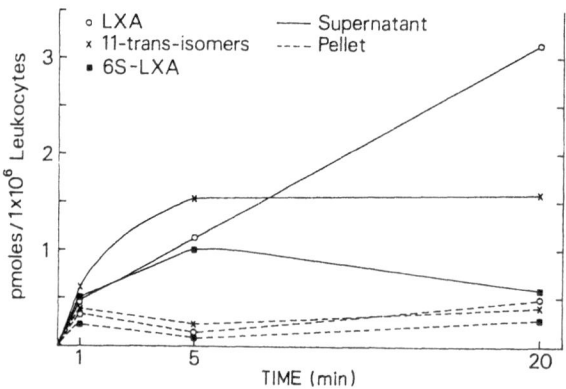

Fig. 6. Time course of LXA_4 generation by human leukocytes. Cells were incubated with 15-HETE (50 μM) and A23187 (2.5 μM) at 37°C. At the indicated times, the incubation mixtures were rapidly centrifuged and resulting pellets and supernatants were isolated and extracted (n=3). The relative amounts of LXA_4, 6S-LXA_4 and the 11-trans-isomers are shown. At 60 seconds, approximately equal amounts of LXA_4 are associated with cell pellets (- - -) as with supernatant (———). At times greater than 5 min LXA_4 is accumulated in the supernatant.

rich granulocyte suspensions were obtained from three donor groups which included eosinophilia due to an allergic disorder, reactions to drugs, and hypereosinophilic syndrome. In each case, eosinophils from these patient categories generated leukotriene C_4 in amounts 20-50 times greater than LXA_4 from endogenous sources of arachidonic acid when exposed to the ionophore A23187[23]. The values obtained for LXA_4 are expressed per 30 x 10^6 leukocytes per incubation, since this was the mean value of cells (or total cell number) obtained following isolation from whole blood. The amount of LXA$_4$ detected in these incubations correlated with the percentage of eosinophils in the leukocyte suspension. This finding suggests that the formation of LXA$_4$ from endogenous sources in eosinophil-rich leukocyte suspensions, exposed to ionophore A23187, is dependent upon the percentage of eosinophils present. It cannot be concluded from these results whether

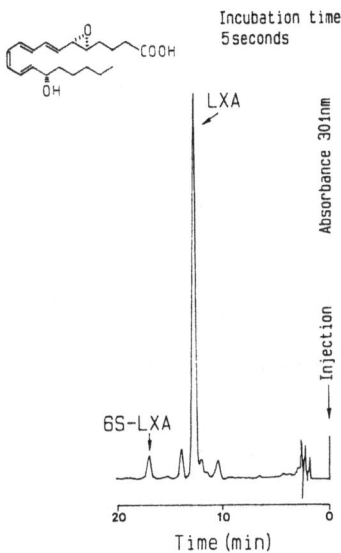

Fig. 7. A model for the enzymatic formation of LXA$_4$. A representative tracing illustrating the transformation of a synthetic 5(6)-epoxide tetraene by a cytosolic epoxide hydrolase. Virtually identical results were found with human or mouse liver cytosolic epoxide hydrolase. Incubation time for the 5(6)-epoxide tetraene with the enzyme was 5 mins. The incubations were stopped and materials extracted[14]. RP-HPLC was performed on an Altex-Ultrasphere-ODS system.

LXA$_4$ was formed exclusively from the arachidonic acid released and processed within a single cell. For example, transcellular metabolism or cell-cell interactions can contribute to the formation of various lipoxygenase-derived products. In particular, activated leukocytes can utilize exogenous 15-HETE to generate both lipoxin A$_4$ and B$_4$. It remains possible, then, that upon addition of A23187 to eosinophil-enriched cell suspensions suitable substrate(s) can be mobilized by one cell or cell type which can be utilized

9

and transformed to LXA$_4$ by another. Nevertheless, taken together, these
results suggest that lipoxin A$_4$ may be involved in events mediated by
eosinophils or that lipoxin A$_4$ may serve some function within these cells.

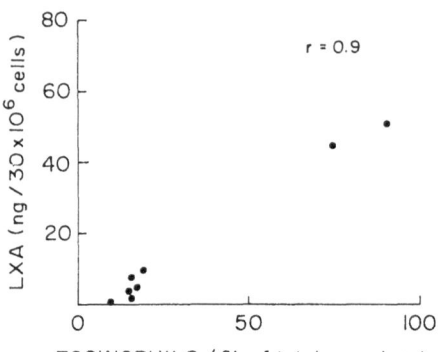

EOSINOPHILS (% of total granulocytes)

Fig. 8. Relationship between amounts of LXA$_4$ formed by granulocytes
from eosinophilic donors following incubation with ionophore
A23187 (2.5 μM) for 20 min at 37°C and the number of
eosinophils (% of total cell suspension) (r = 0.9). Each
point represents a separate cell purification followed by in
vitro challenge. Values are expressed as LXA ng/30 x 10^6
cells/incubation.

The Actions of Lipoxins

To date, the activities of lipoxins have been evaluated in only a few
systems (Table 1). In each system, the responses evoked by these compounds
have proven to be highly stereospecific. For example, when added to human
neutrophils, LXA$_4$, but not its 6S isomer (6S-LXA$_4$), provokes chemotaxis
without provoking aggregation or degranulation. In addition, LXA$_4$ promotes
chemokinesis of human neutrophils[7,24]. Further studies revealed that LXA$_4$
also possesses spasmogenic activities. Lipoxin A$_4$, in submicromolar
concentrations, elicited long-lasting contractions of the guinea pig lung
strip[12,25,26]. The response to LXA$_4$ was not due to release of either
acetylcholine, histamine, noradrenaline or cyclooxygenase products. Here,
synthetic LXA$_4$ causes responses which are indistinguishable from those
obtained with leukocyte derived materials and the 6(S) isomer of LXA$_4$
(6(S)-LXA) did not provoke contractions[12,25].

In the guinea pig ileum, unlike LTC$_4$, LXA$_4$ does not stimulate
contraction which suggests a tissue selectivity associated with lipoxin-
induced responses[25]. Intravital microscopy of the hamster cheek pouch
revealed that LXA$_4$ provokes arteriolar dilation but has no visible effects

Table 1. Actions of Lipoxin A_4 and Lipoxin B_4

Cell type or bioassay system	Responses	Reference
Lipoxin A_4:		
Human neutrophil	Chemotaxis without aggregation	Serhan et al. (7) Palmblad et al. (24)
Human neutrophil	Chemokinesis	Lee and Spur (31)
Guinea pig lung strips	Contraction (cyclooxygenase-independent) Thromboxane production (25) Blocks leukotriene C_4-induced contractions	Dahlen et al. (25,26) Lee and Spur (31)
Hamster cheek pouch	Arteriolar dilation without vascular permeability or leukocyte adherence	Dahlen et al. (26)
Systemic and renal hemodynamics in rats	Selective fall in afferent arteriolar resistance, glomerular hyperperfusion, hypertension, and hyperfiltration	Badr et al. (27)
Protein kinase C (PKC) derived from human placental cytosol	Activates (PKC) in the presence of Ca^{2+} and selects substrate specificity	Hansson et al. (29)
Human leukocytes	Augments thromboxane formation	Conti et al. (30)
Human natural killer (NK) cells	Blocks cytotoxicity and orientation of NK golgi towards target cells (K562) by cAMP-independent mechanism(s) which do not alter target-NK cell binding	Ramstedt et al. (8,28)
Lipoxin B_4:		
Human natural killer (NK) cells	Blocks cytotoxicity	Ramstedt et al. (8)

11

on microvascular permeability or leukocyte adherence to venular endothe-lium[26]. When the systemic and renal hemodynamic responses to intrarenal arterial administration of LXA$_4$ were examined in rats, LXA$_4$ induced glomerular hyperperfusion, hypertension and hyperfiltration. In these experiments, micropuncture measurements of glomerular dynamics revealed that LXA$_4$ provoked a selective and dramatic fall in afferent, but not efferent, arteriolar resistance[27]. Taken together, LXA$_4$ displays a pattern of activity in spasmogenic and in microvasculature assays that is distinct from those known for either prostaglandins, thromboxanes or leukotrienes and appears to be highly tissue-specific. In addition, recent studies indicate that LXA$_4$ can provoke the formation of thromboxane[25,30] suggesting that LXA$_4$ can stimulate the release and transformation of arachidonic acid in various tissues and cell types. These observations suggest that lipoxins can regulate the production of cyclooxygenase products. Of most interest, prior exposure of various tissues to LXA$_4$ blocks the tissue's response to leukotriene C$_4$[25]. Taken together, the results to date suggest that lipoxins may represent products formed by lipoxygenase(s) which can regulate both the actions of leukotrienes and the production of cyclooxygenase pathway products. These observations may have wide implications in the events associated with inflammation.

Since results of several studies suggest a role for lipoxygenase products in the immune response and in modulating natural killer (NK) cell function[8,28], we also examined the effects of a variety of lipoxygenase products on natural killer cells. Here, LXA$_4$ and LXB$_4$ blocked NK cell activity against K562 target cells[8]. Neither compound elevated intra-cellular cAMP nor inhibited target cell binding. When tested in parallel, neither 15-HETE, LTB$_4$ nor LTC$_4$ affected NK cell cytotoxicity against the K562 cell line over a similar dose range. Further studies with human NK cells suggest that LXA stereospecifically blocks NK cytotoxicity by disrupting "signals" involved in the orientation of the Golgi which appears to be of importance in cytotoxicity[28].

Given the vast complexity of stimulus-response coupling and the putative roles of eicosanoids in these events, it became of interest to determine whether eicosanoids themselves can serve as intracellular messengers (i.e. modulate enzyme systems within their cells of origin). We therefore examined the action of lipoxygenase products on the activities of isolated preparations of protein kinase C[29]. Here, LXA$_4$ activated the kinase and proved to be more potent than both diacylglyceride (a proposed intracellular signal in the activation of protein kinase C) and native arachidonic acid. A variety of other oxygenated derivatives of arachidonic acid, including leukotriene B$_4$ (LTB$_4$), were without a direct effect on this enzyme. In this isolated enzyme system, LXB$_4$ was found to be approximately 10 times less potent than LXA$_4$ while both 8-trans-LXB$_4$ and 14S-LXB$_4$ were essentially inactive. Results of these studies also indicated that the substrate specificity of the kinase can be modulated depending upon the stereochemistry of the activator. Together, these studies suggest that lipoxygenase products, in particular LXA$_4$, may serve an intracellular role. Whether lipoxins exert their major role intracellularly within their cell type of origin rather than as extracellular signals remains to be determined.

CONCLUSION

In conclusion, results of the present studies suggest a role for interactions among major lipoxygenases and the products formed thereof (viz. lipoxins) in regulating specific cellular responses. In addition, they indicate that the multiple metabolic fates of lipoxygenase products must be considered when examining either 1) the impact of drugs on the release and oxygenation of fatty acids or 2) the action of specific stimuli on the formation of oxygenation products of such fatty acids. Moreover, they provide additional or alternative means by which the oxygenation of arachidonic acid either within various cells or by transcellular metabolism can exert an effect in allergic reactions, inflammation, thrombosis, and host defense. Further studies on the activities of these compounds as well as on the complex enzyme systems involved in their biosynthesis may lead to the development of new therapeutic agents.

ACKNOWLEDGMENTS

Dr. C.N. Serhan was a visiting scientist at the Karolinska Institutet when the initial work described here was carried out. This work was supported by the Swedish Medical Research Council (Project no. 03X-217; to B.S.) and aided by grant 13-506-867 from the American Heart Association, Massachusetts Affiliate, Inc. (to C.N.S.) C.N. Serhan is a recipient of the J.V. Satterfield Arthritis Investigator Award from the Arthritis Foundation.

REFERENCES

1. B. Samuelsson, Leukotrienes: mediators of immediate hypersensitivity reactions and inflammation, Science 220:568 (1983).
2. B. Samuelsson, S. Hammarstrom, M. Hamberg, and C.N. Serhan, Structural determination of leukotrienes and lipoxins, in: "Advances in Prostaglandin, Thromboxane and Leukotriene Research," vol. 14, J.E. Pike and D.R. Morton, eds., Raven Press, New York (1985).
3. M. Hamberg, P. Hedqvist, and K. Radegran, Identification of 15-hydroxy-5,8,11,13-eicosatetraenoic acid (15-HETE) as a major metabolite of arachidonic acid in human lung, Acta Physiol Scand. 110:219 (1980).
4. J.J. Murray, A.B. Tonnel, A.R. Brash, L.J. Roberts II, P. Gosset, R. Workman, A. Capron, and J.A. Oates, Release of prostaglandin D_2 into human airways during acute antigen challenge, N Engl J Med. 315:800 (1986).
5. P. Needleman, J. Turk, B.A. Jakschik, A.R. Morrison, and J.B. Lefkowith, Arachidonic acid metabolism, Ann Rev Biochem. 55:69 (1986).
6. C.N. Serhan, M. Hamberg, and B. Samuelsson, Trihydroxytetraenes: A novel series of compounds formed from arachidonic acid in human leukocytes, Biochem Biophys Res Comm. 118:943 (1984).
7. C.N. Serhan, M. Hamberg, and B. Samuelsson, Lipoxins: Novel series of biologically active compounds formed from arachidonic acid in human leukocytes, Proc Natl Acad Sci USA. 81:5335 (1984).
8. U. Ramstedt, J. Ng, H. Wigzell, C.N. Serhan, and B. Samuelsson, Action of novel eicosanoids lipoxin A and B on human natural killer cell cytotoxicity: Effects on intracellular cAMP and target cell binding. J Immunol. 135:3434 (1985).
9. C.N. Serhan, M. Hamberg, and B. Samuelsson, Novel mechanisms in the arachidonic acid cascade: Formation of lipoxins, in: "Advances in Inflammation Research," vol. 10, F. Russo-Marie, ed., Raven Press, New York (1985).
10. C.N. Serhan, P. Fahlstadius, S.-E. Dahlen, M. Hamberg, and B. Samuelsson, On the biosynthesis and biological activities of lipoxins, in: "Advances in Inflammation Research," vol. 15, O. Hayaishi and S. Yamamoto, eds., Raven Press, New York (1985).
11. C.N. Serhan, M. Hamberg, B. Samuelsson, J. Morris, and J. Wishka, On the stereochemistry and biosynthesis of lipoxin B, Proc Natl Acad Sci USA. 83:1983 (1986).
12. C.N. Serhan, K.C. Nicolaou, S.E. Webber, C.A. Veale, S.E. Dahlen, T.J. Puustinen, and B. Samuelsson, Lipoxin A: Stereochemistry and biosynthesis, J Biol Chem. 261:16340 (1986).

13. C.N. Serhan, K.C. Nicolaou, S.E. Webber, C.A. Veale, J. Haeggstrom, T.J. Puustinen, and B. Samuelsson, Stereochemistry and biosynthesis of lipoxins, in: "Advances in Prostaglandin, Thromboxane and Leukotriene Research," B. Samuelsson, ed., Raven Press, New York (in press).

14. T. Puustinen, S.E. Webber, K.C. Nicolaou, J. Haeggström, C.N. Serhan, and B. Samuelsson, Evidence for a 5(6)-epoxy tetraene intermediate in the biosynthesis of lipoxins in human leukocytes, FEBS Lett. 207:127 (1986).

15. H. Kühn, R. Wiesner, and H. Stender, The formation of products containing a conjugated tetraenoic system by pure reticulocyte lipoxygenase, FEBS Lett. 177:255 (1984).

16. E.J. Corey and M.M. Mehrotra, A stereoselective and practical synthesis of 5,6(S,S)-epoxy-15(S)-hydroxy-7(E),9(E),11(Z),13(E)-eicosatetra-enoic acid (4), possible precursor of the lipoxins, Tetrahedron Lett. 27:5173 (1986).

17. E.J. Corey and W. Su, Simple synthesis and assignment of stereochem-istry of lipoxin A, Tetrahedron Lett. 26:281 (1985).

18. B.J. Fitzsimmons, J. Adams, J.F. Evans, Y. Leblanc, and J. Rokach, The lipoxins: Stereochemical identification and determination of their biosynthesis, J Biol Chem. 260:13008 (1985).

19. J. Adams, B.J. Fitzsimmons, Y. Girard, Y. Leblanc, J.F. Evans, and J. Rokach, Enantiospecific and stereospecific synthesis of lipoxin A. Stereochemical assignment of the natural lipoxin A and its possible biosynthesis, J Am Chem Soc. 107:464 (1985).

20. E.J. Corey, M.M. Mehrotra, and W. Su, On the synthesis and structure of lipoxin B, Tetrahedron Lett. 26:1919 (1985).

21. P.Y.-K. Wong, R. Huyes, and B. Lam, Lipoxene: A new group of trihy-droxypentaenes of EPA derived from porcine leukocytes, Biochem Biophys Res Comm. 126:763 (1985).

22. C.N. Serhan, P.Y.-K. Wong, and B. Samuelsson, Nomenclature of lipoxins and related compounds derived from arachidonic acid and eicosapenta-enoic acid, submitted.

23. C.N. Serhan, U. Hirsch, J. Palmblad, and B. Samuelsson, Formation of lipoxin A by granulocytes from eosinophilic donors, FEBS Lett., in press.

24. J. Palmblad, H. Gyllenhammer, B. Ringertz, C.N. Serhan, B. Samuelsson, and K.C. Nicolaou, The effects of lipoxin A and lipoxin B on functional responses of human granulocytes, Biochem Biophys Res Comm., in press.

25. S.-E. Dahlén et al., in: "Lipoxins: Biosynthesis and Pharmacology", at FASEB, 71st annual meeting, Washington, D.C., March, 1987.

26. S.-E. Dahlen, J. Raud, C.N. Serhan, J. Bjork, and B. Samuelsson, Biological activities of lipoxin A include lung strip contraction and dilation of arteriols in vivo, Acta Physiol Scand., submitted.

27. K. Badr, C.N. Serhan, K.C. Nicolaou, and B. Samuelsson, Action of lipoxin A on glomerular microcirculatory dynamics in the rat, Biochem Biophys Res Comm., in press.

28. U. Ramstedt, C.N. Serhan, K.C. Nicolaou, S.E. Webber, H. Wigzell, and B. Samuelsson, Lipoxin A-induced inhibition of natural killer cells: Studies on stereospecificity and mode of action, J Immunol. 1:266 (1987).

29. A. Hansson, C.N. Serhan, M. Ingelman-Sundberg, and B. Samuelsson, Activation of protein kinase C by lipoxin A and other eicosanoids: Intracellular action of oxygenation products of arachidonic acid, Biochem Biophys Res Comm. 134:1215 (1985).

30. P. Conti, M. Reale, A. Cancelli, and P.U. Angeletti, Lipoxin A augments release of thromboxane from human polymorphonuclear leukocyte suspensions, FEBS Lett., in press.

31. T.H. Lee and B. Spur, in: "Lipoxins: Biosynthesis and Pharmacology", at FASEB, 71st annual meeting, Washington, D.C., March, 1987.

LIPOXIN SYNTHESES BY ARACHIDONATE 12- AND 5-LIPOXYGENASES PURIFIED FROM PORCINE LEUKOCYTES

Shozo Yamamoto[1], Natsuo Ueda[1], Chieko Yokoyama[1],
Brian J. Fitzsimmons[2], Joshua Rokach[2], John A. Oates[3],
and Alan R. Brash[3]

[1]Department of Biochemistry, Tokushima University
 School of Medicine, Tokushima 770, Japan
[2]Merck Frosst Canada, Inc., Pointe Claire-Dorval
 Quebec, Canada H9R 4P8
[3]Department of Pharmacology, Vanderbilt University
 School of Medicine, Nashville, Tennessee 37232, U.S.A.

Recently, we were successful in the purification of 12-lipoxygenase[1] and 5-lipoxygenase[2] from porcine leukocytes by immunoaffinity chromatography using their monoclonal antibodies which were raised with the crude enzyme preparations as antigen[3,4]. In this chapter the lipoxin biosynthesis will be discussed as catalytic properties of these purified lipoxygenases.

LIPOXIN SYNTHESES BY 12-LIPOXYGENASE

Starting with the cytosol of porcine leukocytes, 12-lipoxygenase was partially purified[5], and then applied to immunoaffinity chromatography using a species of monoclonal anti-12-lipoxygenase antibody which was immobilized to Affi-gel 10[1]. A bulk of protein passed through the column, and 12-lipoxygenase was eluted by increasing the pH and the concentration of detergent and sodium chloride. A near homogeneity of the purified enzyme was demonstrated by SDS-polyacrylamide gel electrophoresis and silver staining. The average specific activity of the purified enzyme was about 4 μmol/min/mg protein at 24°C[1]. As shown in Fig. 1, the purified 12-lipoxygenase reacted with various substrates in addition to arachidonic acid[1]. Unlike 5-lipoxygenase to be discussed later, 12-lipoxygenase was inactive with 12-HPETE which was the primary product from arachidonic acid. 5-HPETE and 5-HETE underwent 12-oxygenation. An important finding was the reactivity of the enzyme with 15-HPETE. The

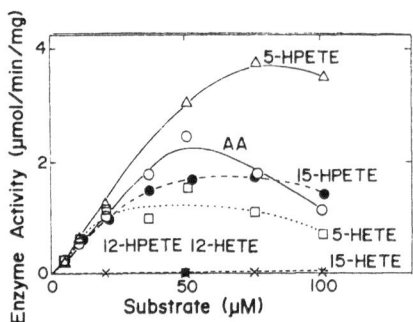

Fig. 1. Reactivities of 12-lipoxygenase with various substrates.

reaction occurred at a rate comparable with the rate of arachidonate 12-oxygenation.

In order to find out what were made from the 15-HPETE, we incubated the enzyme with ^{14}C-labeled 15-HPETE and separated the products by thin layer chromatography (Fig. 2). Several polar products were found in the reaction of a native enzyme but essentially none in the incubation with a heat-denatured enzyme. The reaction products were reduced by the addition of sodium borohydride, and analyzed by straight-phase HPLC, as shown in the upper panel of Fig. 3. Each absorption peak monitored at 270 nm χ for a conjugated triene was compared with an authentic compound previously prepared and identified by the Vanderbilt group[6,7]. Four major peaks were observed. The compound coming out as the highest peak was identified as 14R,15S-dihydroxy acid with 8-cis,10-trans,12-trans-triene (14R,15S-diHETE). Its 14-epimer was found in a negligible amount. The following small peak was associated with 8S,15S-dihydroxy acid with 9-

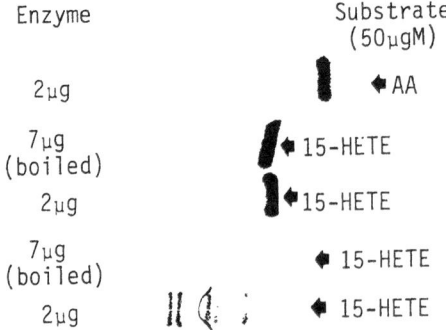

Fig. 2. Reaction products from 15-HPETE by 12-lipoxygenase as analyzed by thin layer chromatography.

16

trans,11-cis,13-trans-triene (8S,15S-diHETE), which was followed by a minor peak of its 8-epimer. Then, two peaks of almost the same height were eluted. They were 8R,15S-dihydroxy acid with all-trans-triene and its 8-epimer. The middle panel presents a profile of the reaction products which were extracted without borohydride reduction. No peaks appeared with a retention time of 14R,15S-diHETE and 8S,15S-diHETE, respectively and two faster migrating peaks were observed, while the retention time of the other peaks remained unchanged. Furthermore, as

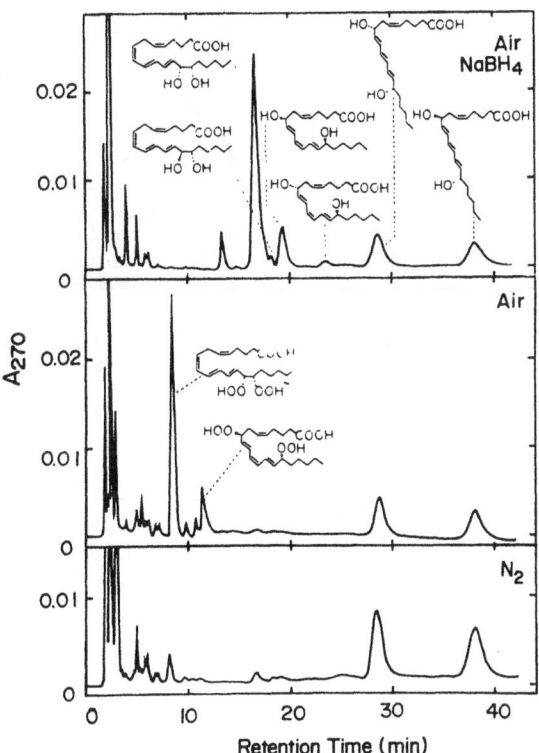

Fig. 3. Straight-phase HPLC of reaction products from 15-HPETE
        by 12-lipoxygenase. (upper) Aerobic reaction products
        were reduced with sodium borohydride. (middle) Aerobic
        reaction products were not reduced. (lower) Anaerobic
        reaction products were not reduced.

shown in the lower panel, the two fast peaks were hardly detected in the absence of air while the production of the other two peaks was not affected by the presence or absence of air. These findings suggested that the fast migrating two peaks were 14R,15S- and 8S,15S-dihydroperoxy acids. Fig. 4 illustrates how these compounds are produced from 15-HPETE by the catalyses of 12-lipoxygenase. First, the enzyme funcitons as

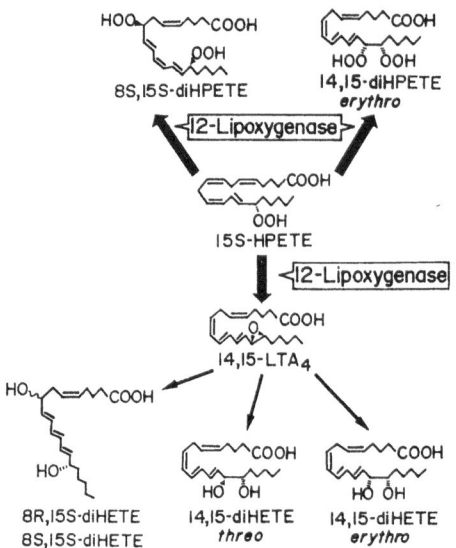

Fig. 4. Reactions catalyzed by 12-lipoxygenase with 15-HPETE as
substrate.

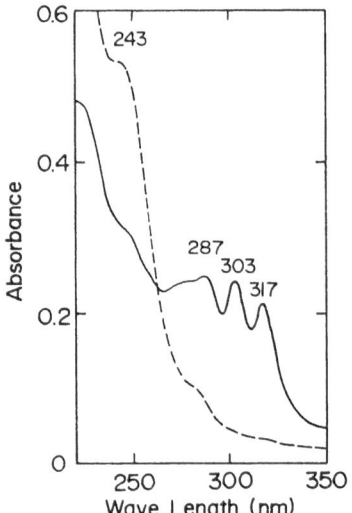

Fig. 5. UV absorption spectrum of the reaction products from
5,15-diHPETE by 12-lipoxygenase.

14R-oxygenase or 8S-oxygenase with 15-HPETE as substrate, and aerobically produces corresponding dihydroperoxy acids. Secondly, the 15-HPETE is anaerobically transformed to 14,15-leukotriene A₄. The epoxy product is readily hydrolyzed non-enzymatically and converted to a racemic mixture of 8,15-diHETE and 14,15-diHETE, respectively. Thus, 12-lipoxygenase shows both the oxygenase and leukotriene A synthase activities.

In consideration of these activities of 12-lipoxygenase on one hand and the chemical structure of lipoxins A[8] and B[9] on the other hand, we examined a possible production of lipoxins by the reaction of 12-lipoxygenase with 5,15-diHPETE. When the purified 12-lipoxygenase was incubated with 5,15-diHPETE and the reaction mixture was reduced by the addition of borohydride, we observed an absorption spectrum with maximum at 303 nm and shoulders around 287 and 317 nm (Fig. 5). Such a spectrum suggested the production of compounds with a conjugated tetraene. As shown in Fig. 6, the reaction products were analyzed by reverse-phase HPLC monitoring absorption at 301 nm. Several peaks were observed and identified by comparison with authentic lipoxin isomers synthesized by the Merck Frosst group[10]. The highest peak comigrated with 5S,14R,15S-trihydroxy acid with 6-trans,8-cis,10-trans,12-trans-tetraene, which was referred to as lipoxin B by the Karolinska group[9]. Since the peak height was markedly reduced in the absence of air, the primary product was oxygen-dependent and presumably 5S,14R,15S-trihydroperoxy acid. The doublet coming out

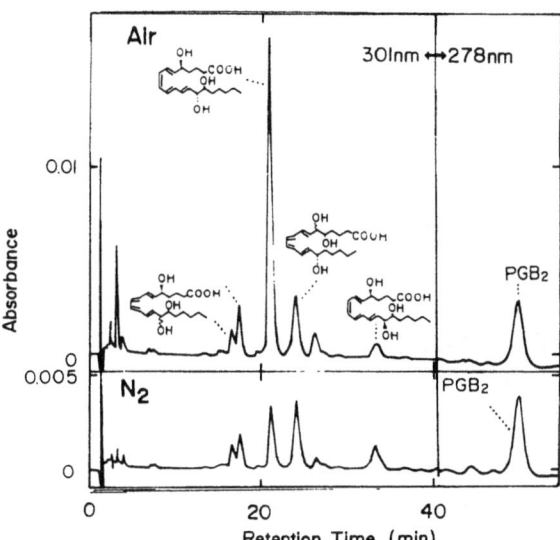

Fig. 6. Reverse-phase HPLC of reaction products from 5,15-diHPETE by 12-lipoxygenase. (upper) Aerobic and (lower) anaerobic reaction products were reduced with sodium borohydride.

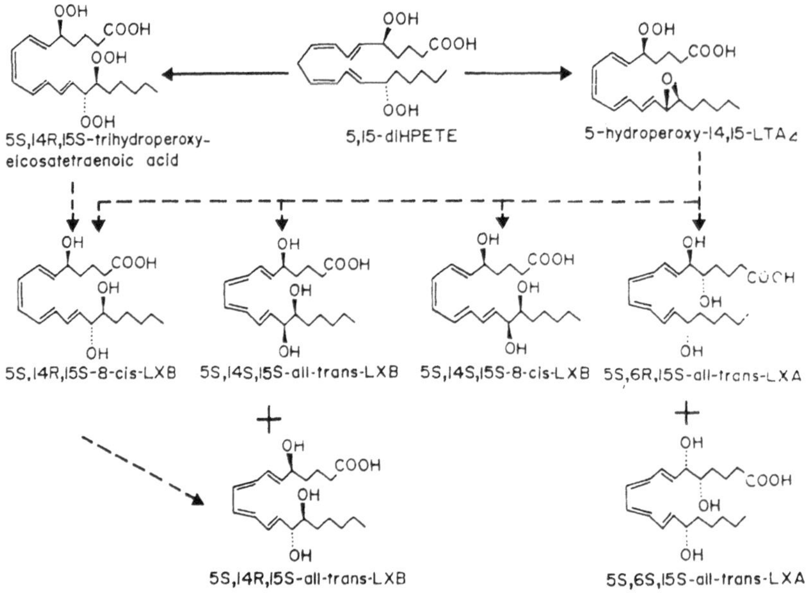

Fig. 7. Pathways of lipoxin syntheses catalyzed by 12-
lipoxygenase.

faster than the highest peak, corresponded to 5S,14S,15S-trihydroxy acid
and its 14-epimer both with all-trans tetraene. A peak a little behind
the highest peak was associated with a mixture of 5S,6S,15S-trihydroxy
acid and its 6-epimer with all-trans tetraene. Another peak eluted far
behind the highest peak was identified as the 14-epimer of lipoxin B.
In contrast to the highest peak, the other peaks were produced almost in
the same amounts both in the presence and absence of air.

Formation of these compounds from 5,15-diHPETE may be rationalized
as illustrated in Fig. 7. As discussed above for 15-HPETE, the 12-lipo-
xygenase also catalyzes the 14R-oxygenation of 5,15-diHPETE. The product
is reduced to lipoxin B. Concomitantly, 5,15-diHPETE is transformed to
5S-hydroperoxy-14,15-epoxy compound of leukotriene A-type. This compound
is non-enzymatically hydrolyzed to various lipoxin A and B isomers. In
comparison with the arachidonate 12-oxygenation, the lipoxin formation
from 5,15-diHPETE by 12-lipoxygenase was not a minor reaction in terms of
the reaction rate (Fig. 8).

LIPOXIN SYNTHESES BY 5-LIPOXYGENASE

An important physiological role of 5-lipoxygenase is the biosynthesis of leukotrienes, which are known as chemical mediators of anaphylaxis and inflammation[11]. The cytosol fraction of porcine leukocytes was

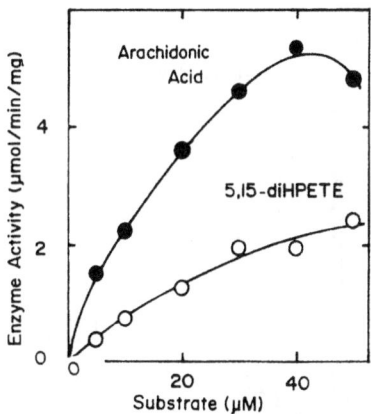

Fig. 8. Reaction rate of 12-lipoxygenase with 5,15-diHPETE
as substrate compared with that of arachidonate
12-oxygenation.

applied to an immunoaffinity chromatographic column using a monoclonal
anti-5-lipoxygenase antibody[2]. The pass-through fraction contained 12-
lipoxygenase together with a bulk of protein. 5-Lipoxygenase preparation
eluted from the column was near homogeneous. The final specific enzyme
activity was about 1 µmol/min/mg protein at 24°C.

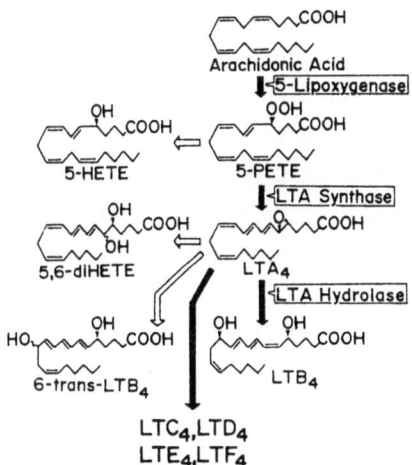

Fig. 9. Reactions catalyzed by 5-lipoxygenase and leukotriene
biosyntheses.

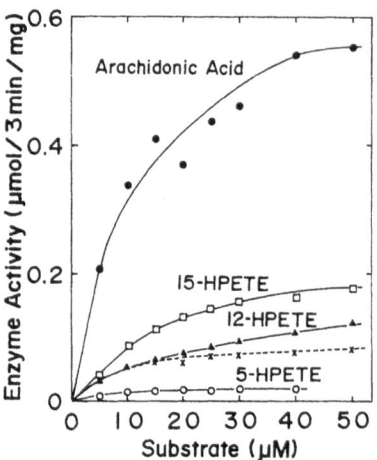

Fig. 10. Reactivities of 5-lipoxygenase with various substrates.

According to the recent studies by several groups[2,12-14], it is now generally accepted that the 5-lipoxygenase is a bifunctional enzyme and a single protein functions as both 5-oxygenase and leukotriene A synthase (Fig. 9). Thus, the enzyme transforms arachidonic acid to 5-HPETE (5-oxygenation) and then the latter compound was converted to leukotriene A_4 with 5,6-epoxide (dehydration). However, as shown in Fig. 10, the leukotriene synthase was only about 6% of the 5-oxygenase in terms of the maximal velocity. Furthermore, 15-HPETE was oxygenated at position-5 at a considerable reaction rate producing 5,15-diHPETE. We examined a possibility if lipoxins were produced from 5,15-diHPETE by the leukotriene A synthase activity of 5-lipoxygenase as discussed for 12-lipoxygenase.

5,15-diHPETE was allowed to react with the purified 5-lipoxygenase[15]. After reaction at 24°C for 5 min, sodium borohydride was added to the reaction mixture, and the extract with ethyl ether was analyzed by reverse-phase HPLC. Absorption at 301 nm was monitored for a conjugated tetraene (Fig. 11). As shown in the upper panel, several peaks were found and compared with authentic lipoxin isomers[10]. The fast migrating doublet corresponded to 5S,14S,15S-trihydroxy acid and its 14-epimer with all-trans tetraene. The following high peak was a mixture of 5S,6S,15S-trihydroxy acid and its 6-epimer both having all-trans tetraene. The next peak comigrated with 5S,6R,15S-trihydroxy acid with 7-trans,9-trans,

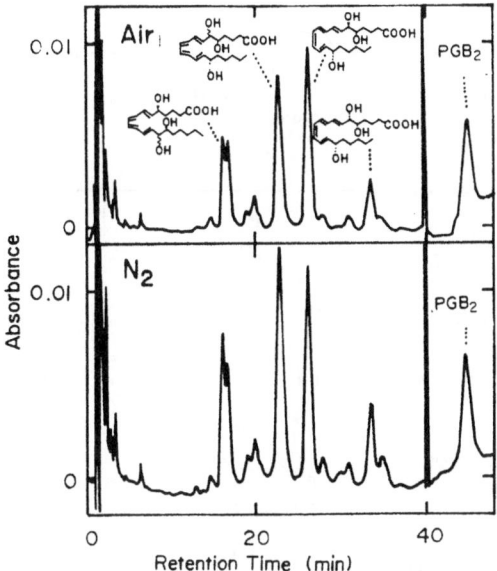

Fig. 11. Reverse-phase HPLC of reaction products from
5,15-diHPETE by 5-lipoxygenase. (upper) Aerobic and
(lower) anaerobic reaction products were reduced with
sodium borohydride.

11-cis,13-trans-tetraene. This compound was designated as lipoxin A by
the Karolinska group[8]. The slow migrating peak was its 6-epimer. When
the reaction was performed in the absence of air, the product profile

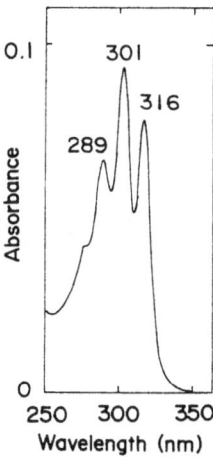

Fig. 12. UV absorption spectrum of lipoxin A produced from
15-HPETE by 5-lipoxygenase.

23

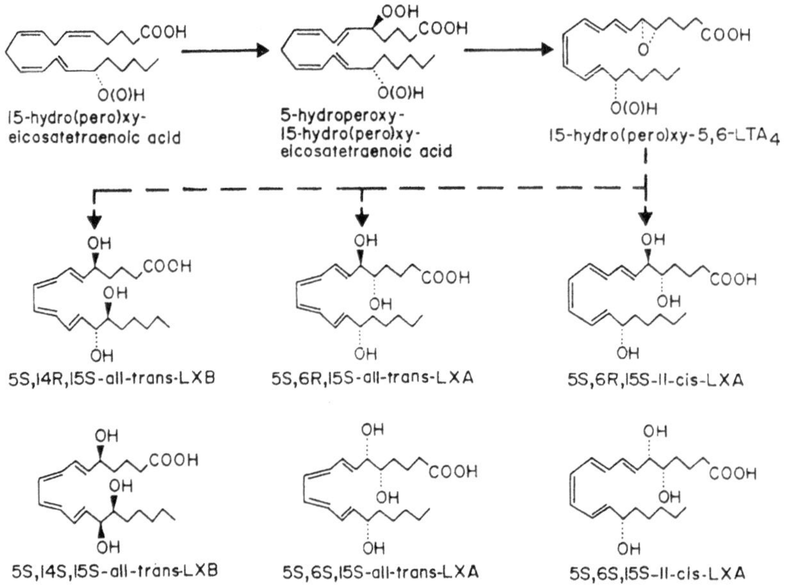

Fig. 13. Pathways of lipoxin syntheses catalyzed by
5-lipoxygenase.

was essentially unchanged. The finding indicated that the production of various lipoxin isomers by 5-lipoxygenase was mostly attributed to the anaerobic epoxide pathway and there was only a minor contribution, if any, of the aerobic multiple oxygenation pathway. Fig. 12 presents a spectrum of the lipoxin A produced by 5-lipoxygenase. The spectrum was indistinguishable from that of authentic lipoxin A with absorption maximum at 301 nm and shoulders at 289 and 316 nm.

Fig. 13 presents a rationale of the lipoxin syntheses by the catalysis of 5-lipoxygenase. 15-HPETE is a substrate of 5-oxygenase, and converted to 5,15-diHPETE, which is further transformed anaerobically to 15-hydroperoxy-5,6-epoxy compound by the leukotriene A synthase activity of the enzyme. This compound is hydrolyzed non-enzymatically to various lipoxin A and B isomers. The rate of conversion of 5,15-diHPETE to these lipoxins was estimated to be about 2% of the arachidonate 5-oxygenation. There was only a minor contribution, if any, of the aerobic lipoxin synthesis by multiple oxygenations.

Lipoxin biosynthesis has been studied so far by other investigators with a whole cell suspension of leukocytes starting with 15-hydroperoxy acids[8-10,16]. Our experiments with the purified 12- and 5-lipoxygenases from porcine leukocytes established an enzymatic nature of lipoxin synthesis from 5,15-diHPETE. The lipoxin synthesis was rationalized by the oxygenase activity and the leukotriene A synthase activity of these lipoxygenases which were previously demonstrated with other substrates[1,2]. Our results clearly demonstrated two ways of lipoxin synthesis. One is an aerobic mechanism by multiple enzymatic oxygenations, and the other is an anaerobic mechanism of enzymatic epoxide production followed by its non-enzymatic hydrolysis. Contribution of these two mechanisms must be confirmed quantitatively by the ^{18}O-experiments, the results of which are now being analyzed.

REFERENCES

1. C. Yokoyama, F. Shinjo, T. Yoshimoto, S. Yamamoto, J.A. Oates, and A.R. Brash, Arachidonate 12-lipoxygenase purified from porcine leukocytes by immunoaffinity chromatography and its reactivity with hydroperoxyeicosatetraenoic acids, J. Biol. Chem. 261:16714 (1986).

2. N. Ueda, S. Kaneko, T. Yoshimoto, and S. Yamamoto, Purification of arachidonate 5-lipoxygenase from porcine leukocytes and its reactivity with hydroperoxyeicosatetraenoic acids, J. Biol. Chem. 261:7982 (1986).

3. F. Shinjo, T. Yoshimoto, C. Yokoyama, S. Yamamoto, S. Izumi, N. Komatsu, and K. Watanabe, Studies on porcine arachidonate 12-lipoxygenase using its monoclonal antibodies, J. Biol. Chem. 261:3377 (1986).

4. S. Kaneko, N. Ueda, T. Tonai, T. Maruyama, T. Yoshimoto, and S. Yamamoto, Arachidonate 5-lipoxygenase of porcine leukocytes studied by enzyme immunoassay using monoclonal antibodies, J. Biol. Chem. 262:in press (1987).

5. T. Yoshimoto, Y. Miyamoto, K. Ochi, and S. Yamamoto, Arachidonate 12-lipoxygenase of porcine leukocyte with activity for 5-hydroxyeicosatetraenoic acid, Biochim. Biophys. Acta 713:638 (1982).

6. R. L. Mass, A.R. Brash, and J.A. Oates, A second pathway of leukotriene biosynthesis in porcine leukocytes, Proc. Natl. Acad. Sci. USA 78:5523 (1981).

7. R.L. Mass, and A.R. Brash, Evidence for a lipoxygenase mechanism in the biosynthesis of epoxide and dihydroxy leukotrienes from 15(S)-hydroperoxyicosatetraenoic acid by human platelets and porcine leukocytes, Proc. Natl. Acad. Sci. USA 80:2884 (1983).

8. C.N. Serhan, K.C. Nicolaou, S.E. Webber, C.A. Veale, S.-E. Dahlén, T.J. Puustinen, and B. Samuelsson, Lipoxin A: Stereochemistry and biosynthesis, J. Biol. Chem. 261:16340 (1986).

9. C.N. Serhan, M. Hamberg, B. Samuelsson, J. Morris, and D.G. Wishka, On the stereochemistry and biosynthesis of lipoxin B, Proc. Natl. Acad. Sci. USA 83:1983 (1986).

10. B.J. Fitzsimmons, J. Adams, J.F. Evans, Y. Leblanc, and J. Rokach, The lipoxins: Stereochemical identification and determination of their biosynthesis, J. Biol. Chem. 260:13008 (1985).

11. B. Samuelsson, Leukotrienes: Mediators of immediate hypersensitivity reactions and inflammation, Science 220:568 (1983).

12. T. Shimizu, O. Rådmark, and B. Samuelsson, Enzyme with dual lipoxygenase activities catalyzes leukotriene A_4 synthesis from arachidonic acid, Proc. Natl. Acad. Sci. USA 81:689 (1984).

13. C.A. Rouzer, T. Matsumoto, and B. Samuelsson, Single protein from human leukocytes possesses 5-lipoxygenase and leukotriene A_4 synthase activities, Proc. Natl. Acad. Sci. USA 83:857 (1986).

14. T. Shimizu, T. Izumi, Y. Seyama, K. Tadokoro, O. Rådmark, and B. Samuelsson, Characterization of leukotriene A_4 synthase from murine mast cells: Evidence for its identity to arachidonate 5-lipoxygenase, Proc. Natl. Acad. Sci. USA 83:4175 (1986).

15. N. Ueda, S. Yamamoto, B.J. Fitzsimmons, and J. Rokach, Lipoxin synthesis by arachidonate 5-lipoxygenase purified from porcine leukocytes, Biochem. Biophys. Res. Commun. 144:996 (1987).

16. P.Y-K. Wong, R. Hughes, and B. Lam, Lipoxene: A new group of trihydroxy pentaenes of eicosapentaenoic acid derived from porcine leukocytes, Biochem. Biophys. Res. Commun. 126:763 (1985).

3

PHOSPHOLIPASE A_2 STIMULATED RELEASE OF LIPOXIN B_4 FORMATION FROM

ENDOGENOUS SOURCES OF ARACHIDONIC ACID IN PORCINE LEUKOCYTES

Patrick Y-K.Wong

Department of Pharmacology
New York Medical College
Valhalla, New York   10595

SUMMARY

Incubation of an isoenzyme of phospholipase A_2 (PLA_2, isolated from snake venom) with porcine leukocytes resulted in the formation of several trihydroxytetraene- containing compounds which were derived from endogenous sources of arachidonic acid. The formation of these endogenous compounds was dose-dependent with an EC_{50} of approximately 1.25×10^{-8}M. At this concentration of the isoenzyme and time (10 min) of explosure the cells remained viable as determined by the exclusion of trypan blue. The trihydroxytetraene compounds were purified by RP-HPLC and their identities were analyzed by U.V. spectrometry, GC/MS and by comparison with synthetic materials. The biologically derived compounds proved to be lipoxin B_4 (5S,14R,15S-trihydroxy-6,10,12-trans-8-cis-eicosatetraenoic acid) and its two structural isomers (8-trans-LXB_4 and 14S-8-trans-LXB_4). Results of the present study indicate that porcine leukocytes can generate lipoxin B_4 and its isomers from endogenous sources of arachidonic acid. Moreover, they suggest that certain PLA_2 isoenzymes may stimulate the formation of lipoxins and related compounds.

Abbreviations

15-HPETE	15-hydroperoxyeicosatetraenoic acid
15-HPEPE	15-hydroperoxyeicosapentaenoic acid
AA	Arachidonic Acid
EPA	Eicosapentaenoic acid
RP-HPLC	Reverse Phase-High Performance Liquid Chromatography
GC/MS	Gas Chromatography/Mass Spectrometry
LXB_4	Lipoxin B_4:(5S,14R,15S)-5,14,15-trihydroxy-6,10,12-trans-8-cis-eicosatetraenoic acid
8-trans-LXB_4	(5S,14R,15S)-5,14,15-trihydroxy-6,8,10,12-trans-eicosatetraenoic acid
14S-8-trans LXB_4	(5S,14S,15S)-5,14,15-trihydroxy-6,8,10,12-trans-eicosatetraenoic acid;
PLA_2	Phospholipase A_2

Lipoxin is a new series of trihydroxytetraene metabolites of 15-hydroxyperoxide of AA and EPA from human and porcine leukocytes as reported by Serhan et al. (1,2) and Wong et al. (3). This group of new compounds displays a unique conjugated tetraene ultraviolet spectrum with absorption maxima at 301-302 nm, with shoulders at 287 and 316 nm. Serhan and co-workers reported the isolation of two major isomers, lipoxin A_4 and B_4 from 15-HPETE with the structural assignment as 5,6,15L-trihydroxy-7,9,11,13-eicosatetraenoic acid and 5D,14,15L-trihydroxy-6,8,10,12-eicosatetraenoic acid (1,4). Similarly, 15-HPEPE of EPA was also demonstrated to be the precursor of the Lipoxin A_5 and B_5 (3). The retention of stereochemistry of the hydroxy groups and double bonds further suggested that the formation of lipoxin A_4 and B_4 in human leukocytes was an active enzymic process (5). The biological activity of lipoxin A_4 includes contraction of guinea-pig lung strips at 10^{-7}M and causes the release of a superoxide anion (O_2^-) from human PMNL (2,6). Furthermore, the same concentration of lipoxin A_4 was capable of inhibiting natural killer cell activities (2). In addition, lipoxin A_4 induces contraction of lung parenchymal strips and increases microvascular permeability (7). More recently, lipoxin A_4 (LXA_4) has been found to activate isolated preparations of human placental-derived protein kinase C (in vitro) (8) and proved to be 30 times more potent than diacylglycerol, a proposed intracellular signal for the activation of protein kinase C (9). Like lipoxin A_4, lipoxin A_5 derived from 15-HPEPE also induces superoxide anion generation with similar potency to that of LXA_4 without causing aggregation in canine neutrophils (10). Thus, in addition to 5-lipoxygenase, 15-lipoxygenase metabolites of AA and EPA can also be converted to groups of compounds with potent biological activities.

Since the biological activities of both series of lipoxins are different from those of LT or PGs (10,11,12), it is important to demonstrate their formtion from endogenous sources of arachidonic acid. Recently, we have demonstrated that lipoxins of the 4 series (AA-derived products) and 5 series (EPA-derived products) can be isolated from porcine leukocytes exposed to either AA or EPA, respectively (10). The results of these studies taken together with the stereospecific nature and the biological activities observed with synthetic and authentic lipoxins suggest that these compounds may play an important role in inflammation. In this study, we investigated the possibility that the lipoxins may be generated from endogenous sources following exposure of the cells to appropriate physiological stimuli. Although at the present time, a receptor-mediated activation of the 15-lipoxygenase has not been demonstrated, a number of 15-lipoxygenase pathway products have been isolated, suggesting that the 15-lipoxygenase is a major route of arachidonic acid metabolism in several tissues (13,14). However, the mechanism(s) involving the activation of 15-lipoxygenase remains uncertain.

It is clear from the results of numerous studies that leukocytes can discharge the contents of their lysosomal granules inadvertently into the cell's environment when they encouter phagocytic stimuli (15). The granule associate enzyme can then attack a variety of host-derived substrates. This release of lysosomal constituents from inflammatory cells (macrophages, neutrophils, etc.) may play a central role in inflammation (15). From the studies of Elsback and coworkers (16) it is clear that PLA_2 is one of the many enzymes associated with the granules of polymorphonuclear leukocytes. It has been suggested that following phagocytosis and degranulation, this granule-associated PLA_2 may participate in the intravascular antibacterial assault of leukocytes on ingested organisms (16), and PLA$_2$ activity has been demonstrated in cell-free supernatants from peritoneal exudate cells (17,18). Thus, it has been proposed that antigen-induced release of PLA$_2$ from activated leukocytes may mediate the generation of lipid mediators at sites of chronic inflammation (19,20). Therefore, we sought evidence that extracellular PLA$_2$ may release novel eicosanoids from inflammatory cells. Using an isoenzyme of PLA$_2$ purified from snake venom (Vipera Russelli), we present evidence indicating that this PlA$_2$ provokes the formation of LXB$_4$ from endogenous sources of AA in porcine leukocytes.

METHODS OF PROCEDURE

Phospholipase A$_2$ isoenzyme

Phospholipase A$_2$ isoenzymes were prepared from Vipera Russelli venom (Sigma Chemical, St. Louis, MO), according to the method of Salach et al. (21), by isoelectric focusing and further purified by HPLC using a Protein-PAK DEAE 5 PW column and 20 mM Tris-acetate Buffer, pH 7.8, with 5% glycerol as mobile phase (Waters Assoc., Milford, MA). The isoenzyme with an isoelectric point of 8.8 to 9.0 was used in this study (molecular weight of this isoenzyme was estimated to be 15,000 Dalton as determined by SDS-gel electrophoresis).

Cell Preparation and Incubation

Fresh porcine blood was obtained from a local slaughter house. Blood was collected in a large plastic container containing 10 mM EDTA. The leukocytes were separated from the red blood cells by dextran sedimentation as described by Borgeat et al. (22). Contaminating red blood cells were lysed with 0.75% NH$_4$Cl. The leukocytes were washed twice and finally resuspended in Dulbecco's phosphate buffered saline, pH 7.4, and adjusted to 100 x 10^6 cells/ml. The viability of the cells as measured by the trypan blue exclusion test was found to be greater than 95%. This cell preparation was contaminated with platelets and mononuclear leukocytes (22). Cell suspensions (50-100 ml) were added to the incubation vessels and incubated at 37°C with constant shaking. The incubation was terminated by the addition of two volumes of ethanol.

Extraction and Purification

The incubation precipitate was filtered, and the ethanolic filtrate was evaporated to dryness. The residue was dissolved in 5 ml of distilled water and acidified to pH 4.0 with 1N HCl. The solution was then extracted with 10 volumes of ethyl acetate. the ethyl acetate fraction was rotor evaporated to dryness under vacuum. The residue was dissolved in 200 μl of solvent A and separated by RP-HPLC on a Water's associates Dual Pump system equipped with an RP Ultrasphere ODS column (C$_{18}$-ODS, 5μ, 10 mm x 25 cm, Beckman, Palo Alto, CA), a U=6K injector and λmax variable wavelength detector. The products were eluted with a linear gradient of methanol/water/acetic acid (50:50:05, v/v) (solvent A) to methanol (solvent B) for 45 min at a flow rate of 3 ml/min (3). Column effluents were monitored with a Water's Associates 481 λmax variable wavelength detector set at 302 nm (0-12 min) and 237 nm (12-45 min). Fractions of 1 ml were simultaneously collected with an on-line fraction collector and a portion of each fraction was removed for estimation of recovered radioactivity. PGB$_2$ was used as internal standard for di-HETEs. The day-to-day variations in retention times were determined to be less than 5%.

<u>UV Spectroscopy</u>

Samples eluted from the HPLC were rotor evaporated to dryness, dissolved in absolute ethanol and examined with a Hewlett-Packard 8450-A UV/Vis spectrophotometer.

<u>Gas Chromatography-Mass Spectrometry</u> (GC/MS)

The methylesters of the tetraene containing materials (Fig 1A and 1B) were converted to trimethylsilyl ethers by addition of 25 μl of pyridine followed by 50 μl of trimethylchlorosilane and 50 μl of hexamethyldisilazine (Supelco). The mixtures were kept at room temperature for 20 min and dried under N_2. The samples were dissolved in 5 μl hexane and injected into the gas chromatograph-mass spectrometer. GC/MS was performed with a Dani 3800 gas chromatograph HR PRV-2CH equipped with a fused silica capillary column (20m x 0.32, Orion) SE-30 and 7070E VG analytical mass spectrometer. The electron energy was set at 22.5 eV, with an oven temperature of 230°C.

RESULTS

Preliminary studies with crude PLA_2 from snake venom incubated with leukocytes and other cell types indicated that several vasoactive products were generated (Wong et al., unpublished data). Porcine leukocytes generated several tetraene-containing materials following exposure to the isolated PLA_2 (Fig. 1A). The formation of these materials by PLA_2 was dose dependent as indicated in Fig 2A. At the highest dose of PLA_2 (3 x 10^{-7} M), the viability of leukocytes after 10 min of incubation was between 65 and 72% (Fig 2B) (n=3). Following 10 min exposure to 10^{-8}M of PLA_2 the viability of the leukocytes was greater than 90%, and the formation of these materials from these cells was more than one-half of the maximum response (EC_{50}:1.25 x 10^{-8}M) (Fig 2B). This result suggests that the generation of these tetraene-containing compounds is not due to cell injury or cellular toxicity caused by PLA_2.

To further substantiate the identity of this tetraene-containing material, samples eluted from RP-HPLC (fraction A) (Fig 1A) were treated with diazomethane and re-chromatographed on a second RP-HPLC as recently described for the separation of LXB_4 (6,23). As shown in Figure 1B, the material was resolved into three components. These biologically-generated materials co-migrated with authentic standards of lipoxin B_4 (LXB_4), 8-trans-lipoxin B_4 and 14S-8-trans-lipoxin B_4 (23). Interestingly, only small amounts of material co-migrated with authentic standards of lipoxin A_4 and its isomers were detected. Each of the materials eluted under the three peaks was collected separately and their UV spectra were recorded. Each material displayed a UV spectrum typical of a conjugated-tetraene (Fig 1 insert). To further elucidate the structure of these three materials, samples were derivatized with TMS and analyzed by GC/MS.

GC/MS structural analysis was performed on the material eluted at 28.5 mins from RP-HPLC. On GC this compound gave a C value of 24.0, identical to that previously reported for lipoxin B_4 (2,23). Its mass spectrum showed ions of high intensities at m/e 173 (base peak, [Me_3 SiO^+ = CH-$(CH_2)_4$ - CH_3] and 203 [Me_3 SiO^+ = CH-$(CH_2)_3$ - $COOCH_3$]. Ions of lower intensities were observed at m/e 582[M], 492[M-90, loss of Me_3SiOH], 482[M-100], 409[M-173], 379[M-203], 319[409-90] and 289[379-90]. These fragmentation ions and the C value are consistent with those reported for lipoxin B_4 (Fig 3).

GC/MS analysis of material collected under the peak, with retention time of 30.6 min (C value=28.3) showed ions of high intensities at m/e 173 (base peak) and 203. Ions of lower intensities were observed at m/e 582, 492, 482, 409, 379, 319 and 289. These findings were essentially

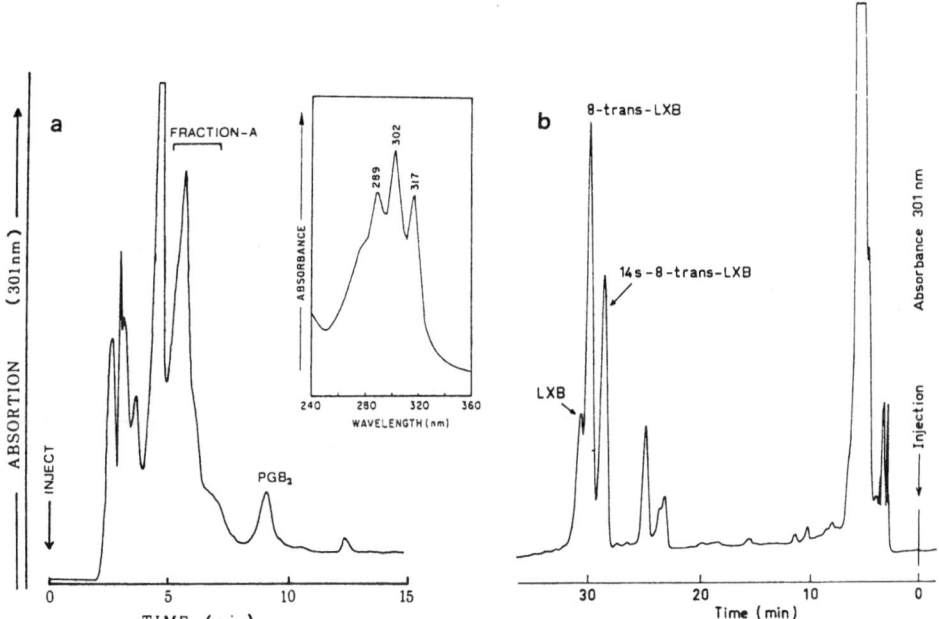

Fig 1. RP-HPLC chromatograms and UV spectrum of products extracted from incubation of porcine leukocytes with PLA_2 isoenzyme (1×10^{-7}M) A) RP-HPLC chromatogram. The column (Beckman, Ultrasphere ODS, 5 μM, 4.6 mm x 2.5 cm) was eluted on a linear gradient of MeOH/$H_2$0/HAC, 50/50/0.05 (V/V) to methanol at 1.0 ml/min; (B) RP-HPLC chromatogram of methylated fraction A. The column was eluted with MeOH/$H_2$0/65/35 (v/v), at a flow rate of 1.0 ml/min. (insert) UV spectrum of materials under Fraction A. The spectrum was recorded in ethanol.

identical to the authentic and reported spectrum of 8-trans-lipoxin B_4 (23). GC/MS analysis of the TMS derivatives of the material eluted at 31.2 min gave a C value of 28.0. Its mass-spectrum showed prominent ions at m/e, 173 [base peak, $Me_3 SiO^+ = CH-(CH_2)_4 - CH_3$], 203[$Me_3 SiO^+ = CH - (CH_2)_3 - COOCH_3$)]. Ions of lower intensities were observed at m/e 582 (M), 492[M-90], 482[M-100; rearrangement followed by loss of O=HC-$(CH_2)_4 - CH_3$), 409[M-173], 379[M-203] and 289[379-90]. The mass spectrum as well as the C value was essentially identical to that reported for 14S-8-trans-lipoxin B_4 (23).

Thus, the tetraene-containing materials formed by porcine leukocytes were identified unequivocally as lipoxin B_4 and its nature stereoisomers.

DISCUSSION

Lipoxins were first reported by Serhan and coworkers and their basic structures determined following incubation of 15-HPETE and human leukocytes (1,2). The stereochemistry of human leukocyte-derived lipoxin A_4 [(5S,6R,15S)-5,6,15-trihydroxy-7,9,13-trans-11 trans- eicosatetraenoic

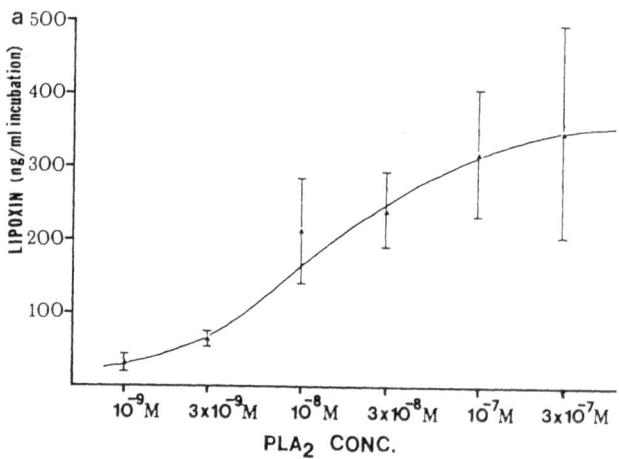

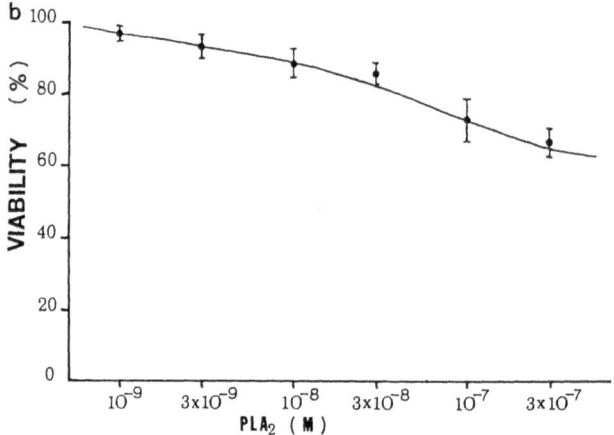

Fig. 2. A) Endogenous release of tetraene-containing materials from porcine leukocytes after stimulation with various doses of PLA_2 isoenzyme. Amount was calculated from UV absorbance using $\bar{E}$=50,000 at 302 nm. Error bar indicates SEM (n=3)
B) Viability of porcine leukocytes after incubation with various doses of PLA_2 isoenzymes for 10 min at 37°C. Viability was determined by exclusion of Trypan blue. Error bar indicates SEM (n=3).

acid) and lipoxin B₄ (5S,14R,15S)-5,14,15-trihydroxy-6,10,12-trans-8-cis-eicosatetraenoic acid)] as well as several of their isomers have been determined (1,2,5). In addition to both AA and 15-HPETE serving as substrate for the formation of LXA₄ and LXB₄, 15-HETE has been found to be transformed to LXA₄ and LXB₄ in activated human leukocytes (5,6,23). The results of these studies suggest that lipoxin can be formed in part by transcellular metabolism of 15-HETE (6,23). It is obvious that several biosynthetic routes could yield tetraene compounds.

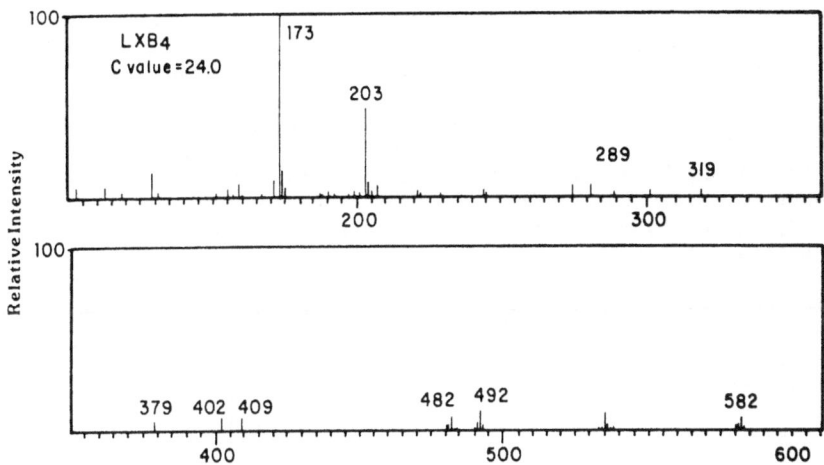

Fig. 3.   Mass spectra of the Me₃Si derivatives of methylesters of materials eluted at retention time 28.5 min.

Thus it remained to be determined whether lipoxins can be formed from endogenous AA released from cellular pools. Recently, we demonstrated that porcine leukocytes incubated with either AA or EPA generated lipoxins (0.05% for AA and 0.1% for EPA) which is about 15-30% of the amount of leukotriene B₄ produced. These results were based on RP-HPLC, UV and GC/MS data (Wong et al., unpublished data). We have hypothesized that lipoxins may also be formed from endogenous AA if appropriate stimuli are applied as in the case of bradykinin stimulation of human platelets to release 15-lipoxygenase products from endogenous sources (24). In this study we have found that Ca⁺⁺ ionophore A23187 (1-10 μM) was a relatively weak stimulus for lipoxin production by porcine leukocytes (less than 50±10 ng of lipoxin B₄ and its isomers were isolated from each 100 x 10⁶ cells incubation, n=8). Bradykinin was not active with the doses tested (up to 10 μM). In contrast, the PLA₂ isoenzyme (pI=8.9) provokes the formation of lipoxin B₄ and its isomers (approximately 30 ng/ml; 100 x 10⁶ cells) from porcine leukocytes at doses as low as 10⁻⁹M.

The rationale for the use of purified PLA$_2$ isoenzyme in this study is based on the following: [1] granule-associated PLA$_2$ of inflammatory cells may be released to the extracellular environment upon cell activation (16-20); [2] that a similar PLA$_2$ isoenzyme has previously been demonstrated to induce leukotriene release in isolated perfused guinea-pig lung (25); [3] we have found that other PLA$_2$ isoenzymes (other than pI of 8.9) isolated from crude Russelli Vipera's venom do not induce the release of lipoxins from porcine leukocytes; finally, [4] we have also found that this isoenzyme provokes the formation of these compounds by certain cell types, i.e., porcine and human leukocytes, rat PMN but not by guinea-pig PMNs, guinea macrophage and rat macrophages (Wong et al, unpublished data). The responses of porcine leukocytes to exogenously added PLA$_2$ as shown in this study may mimick the _in situ_ pathophysiologic response of leukocytes exposed to extracellular PLA$_2$ which may be released by macrophages or leukocytes during cell activation and phagocytosis or in antigen induced chronic inflammation (16,17) (Fig. 4). Furthermore, PLA$_2$ have been reported to be released by rat platelets during platelet aggregation (19) and has been found at the site(s) of inflammation and in lymph draining nodes with tuberculin reactions in rabbits (18). Thus, PLA$_2$ and its isoenzymes released from macrophages, platelets or isolated from snake venom may be useful tools in studying the formation of eicosanoids from endogenous sources during cell activation.

In the present study, the identities of these materials were established by UV, HPLC, and by GC/MS and comparison with synthetic materials. The formation of lipoxin B$_4$ and its isomers provoked by PLA$_2$ was dose dependent (Fig. 2A). Maximal recovery of LXB$_4$ was achieved at about 1×10^{-7}M PLA$_2$. At this dose, approximately 70-80% of leukocytes were viable after 10 min of incubation at 37°C as determined by the Trypan blue exclusion technique (Fig. 2B). Similar results were also obtained from lactate dehydrogenase release assay (data not shown). The selective formation of large amounts of LXB$_4$ but not LXA$_4$, may reflect the involvement of a specific biosynthetic pathway in porcine leukocytes. Previous studies in human leukocytes using 15-HETE suggested that lipoxins can be formed via 5,6-epoxidetetraene intermediates (5,6,23). The high activity of 12-lipoxygenase in porcine leukocytes (26,27) may accelerate formation of lipoxins, since 12-lipoxygenase can also metabolize 15-HPETE to form 14,15 DHETE (26) and may represent another biosynthetic pathway for LXB$_4$ formation. It is of interest to note that purified 12-lipoxygenase from porcine leukocytes has recently been reported to generate LXB$_4$ from 5,15-DHPETE (27). From the results of the presenty study, however, it is not possible to determine the precise route and intermediates involved in the formation of LXB$_4$ and its isomers by porcine leukocytes. Lipoxins have been demonstrated to affect the functions of inflammatory cells, natural killer cells, protein kinase C, and microvascular circulation (2,6,7). Thus, demonstration that lipoxins can be formed from endogenously derived AA (Fig. 1 and 2) in addition to transcellular metabolism of 15-HETE (6,23) provides further evidence that lipoxins may serve as lipid mediators or intracellular regulators in various inflammatory diseases and immune responses (Fig. 4). Moreover, the result of the present study provide further evidence that the release of PLA$_2$ activity by inflammatory cells may result in the formation of agents or mediators which are biologically active in chronic inflammatory processess.

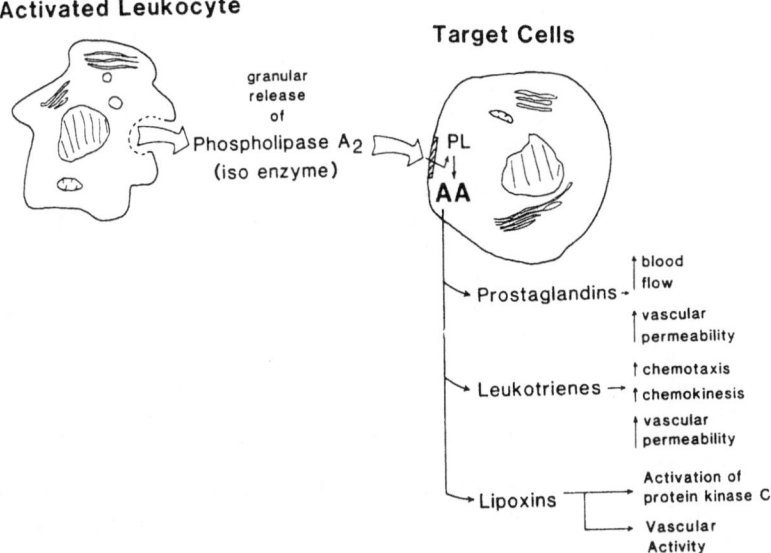

Fig. 4. Proposed mechanism for PLA$_2$ stimulated release of Lipoxin and other lipid mediators from endogenous sources of arachidonic acid in leukocytes.

ACKNOWLEDGEMENTS

This work was supported by National Institute of Health Grant HL-25316. We thank Carl Ehmer Farm and personnel of Poughkeepsie, N.Y. for the collaboration and the generous supply of porcine blood used in this study. The assistance of Gail Price in the preparation of this manuscript is gratefully acknowledged. We also are in debt to Professor B. Samuelsson and Dr. C.N. Serhan for their collaboration during the cause of this study.

REFERENCES

1. Serhan, C.N., Hamberg, M., and Samuelsson, B., Trihydroxytetraenes: A novel series of compounds formed from arachidonic acid in human leukocytes, <u>Biochim. Biophys. Res. Commun.</u>, 118:943-949 (1984).
2. Serhan, C.N., Hamberg, M., and Samuelsson, B., Lipoxins: A novel series of biologically active compounds formed from arachidonic acid in human leukocytes, <u>Proc. Natl. Acad. Sci. U.S.A.</u>, 81:5335-5339 (1984).
3. Wong, P.Y-K., Hughes, R.A., and Lam, B., Lipoxene: A new group of trihydroxy pentaene of eicosapentaenoic acid derived from porcine leukocytes, <u>Biochim. Biophys. Res. Commun.</u>, 126:763-772 (1985).

4. Serhan, C.N., Wong, P.Y-K and Samuelsson, B., Nomenclature of lipoxins and related compounds derived from arachidonic acid and eicosapentaenoic acid, <u>Prostaglandins</u>, in press, 1987.

5. Fitzsimmons, B.J., Adams, J., Evans, J.F., Leblanc, Y., and Rokach, J., Lipoxins: Sterochemical identification and determination of their biosynthesis, <u>J. Biol. Chem.</u>, 260:13008-13012 (1985).

6. Serhan, C.N., Nicolaou, K.C., Webber, S.E., Veale, C.A., Dahlen, S.E., Puustinen, T.J., and Samuelsson, B., Lipoxin A: Stereochemistry and biosynthesis, <u>J. Biol. Chem.</u> 261:16340-16345 (1986).

7. Dahlen, S.E., Serhan, C.N., and Samuelsson, B., <u>Acta. Scand. Physiol.</u> in press (1987).

8. Hansson, A., Serhan, C.N., Haeggstrom, J., Ingelman-Sundberg, M., and Samuelsson, B., Activation of protein kinase C by lipoxin A and other eicosanoids. Intracellular action of oxygenation products of arachidonic acid, <u>Biochem. Biophys. Res. Commun.</u> 134:1215-1222 (1986).

9. Nishizuka, Y., The role of protein kinase C in cell surface signal transduction and tumor promotion, <u>Nature (Lond.)</u> 308:693-698 (1986).

10. Wong, P.Y-K., Spur, B., Hirai, A., Yoshida, S., Tamura, Y., and Lam, B., Biotransformation of arachidonic acid (AA) and eicosapentaenoic acid (EPA) into lipoxins and lipoxenes by porcine leukocytes, <u>Federation Proceedings</u> 45:927 (1986) (Abstract)

11. Von Euler, U.S., On the specific vasodilating and plain muscle stimulating substances from accessory genital glands in man and certain animals (prostaglandin and vesiglandin), <u>J. Physiol.</u> 88:213-234 (1937).

12. Hamberg, M., Svensson, J. and Samuelsson, B., Thromboxanes: a new group of biologically active compounds derived from prostaglandin endoperoxides, <u>Proc. Natl. Acad. Sci. USA</u> 72:2994-2998 (1975).

13. Hunter, J.A., Finkbeiner, W.E., Nadel, J.A., Goetzl, E.J., and Holtzman, M.J., Predominant generation of 15-lipoxygenase metabolites of arachidonic acid by epithelial cells from human trachea, <u>Proc. Natl. Acad. Sci. U.S.A.</u> 82:4633-4637 (1985).

14. Narumiya, S., Salmon, J.A., Cottee, F.H., Weatherley, B.C., and Flower, R.J., Arachidonica acid 15-lipoxygenase from rabbit peritoneal polymorphonuclear leukocytes: partial purification and properties, <u>J. Biol. Chem.</u> 256:9583-9592 (1981).

15. Weissmann, G., Smolen, J.E. and Korchak, H.M., Release of inflammatory mediators from stimulated neutrophils, <u>N. Engl. J. Med.</u> 303:27-34 (1980).

16. Victor, M., Weiss, J., Klempner, M.S. and Elsbach, Phospholipase A_2 activity in the plasma membrane of human polymorphonuclear leukocytes, <u>FEBS LETTERS</u> 136:298-300 (1981).

17. Vadas, P., Wasi, S., Movat, H.Z. and Hay, J.B., Extracellular phospholipase A_2 mediates inflammatory hyperemia, <u>Nature</u> 293:583-585 (1981).

18. Vadas, P. and Hay, J.B., Secretion of a hyperemia-inducing moiety by mitogen on glycogen stimulated mononuclear inflammatory cells of sheep and rabbit, <u>Int. Arch. Allergy Appl. Immunol.</u> 62:142-151 (1980).

19. Vadas, P. and Hay, J.B., The release of phospholipase A_2 from aggregated platelets and stimulated macrophages of sheep, <u>Life Sciences</u> 26:1721-1729 (1980).

20. Vadas, P. and Hay, J.B., The appearance and significance of phospholipase A_2 in lymph draining tuberculin reactions, <u>Am. J. Pathol.</u> 107:285-291 (1982).

21. Salach, J.I., Turini, P., Seng, R., Hauber, J., and Singer, T.P., Phospholipase A of snake venom. I. isolation and molecular properties of isoenzymes from Naja Naja and Vipera Russelli venoms, <u>J. Biol. Chem.</u> 246:331-339 (1971).

22. Borgeat, P., Picard, S., and Vallerland, P., Transformation of arachidonic acid in leukocytes. Isolation and Structural analysis of a novel dihydroxy derivatives, <u>Prostaglandins and Medicine</u> 6:557-570 (1981).

23. Serhan, C.N., Hamberg, M., Samuelsson, B., Morris, J., and Wishka, D.G., On the stereochemistry and biosynthesis of lipoxin B, <u>Proc. Natl. Acad. Sci. U.S.A.</u>, 83:1983-1987 (1986).

24. Wong, P.Y-K., Westlund, P., Hamberg, M., Granstrom, E., Chao, P.H-W, and Samuelsson, B., 15-lipoxygenase in human platelets, <u>J. Biol. Chem.</u> 260:9162-9165 (1985).

25. Huang, H-C, Release of slow reacting substance from the Guinea-pig lung by phospholipase A_2 of Vipera Russelli snake venom, <u>Toxicon.</u> 22:359-372 (1984).

26. Maas, R.L. and Brash, A.R., Evidence for a lipoxygenase mechanism in the biosynthesis of epoxide and dihydroxy leukotrienes from 15(s)-hydroperoxyicosatetraenoic acid by human platelets and porcine leukocytes, <u>Proc. Natl. Acad. Sci. U.S.A.</u> 80:2884-2888 (1983).

27. Yokoyama, C., Yoshimoto, T., Yamamoto, S., Oates, J.A., and Brash, A.R., <u>Proc. 5th International Conference on PGs and Related Compounds.</u> June 2, Florence, Italy, (1986).

4

LIPOXYGENASE CATALYZED OXYGENATION OF HYDROXY FATTY ACIDS

TO LIPOXINS

Hartmut Kühn*, Alan R. Brash*, Rainer Wiesner**,
and Lutz Alder***

*Division of Clinical Pharmacology, Vanderbilt University
Nashville, TN 37232
**Institute for Biochemistry, Humboldt University, 1040 Berlin
Hessische Str. 3-4, German Democratic Republic
***Department of Chemistry, Humboldt University, 1040 Berlin
Hessische Str. 1-2, German Democratic Republic

SUMMARY

The pure lipoxygenases from rabbit reticulocytes and soybeans convert a variety of substrates (arachidonic acid, 15-HPETE, 15-HETE, 5-HETE, various DiHETE isomers) to trihydroxy eicosanoids containing a conjugated tetraene system (lipoxins). In general, the methyl esters are better substrates for lipoxin formation than are the free acids. Lipoxygenase inhibitors (5,8,11,14-eicosatetraynoic acid, nordihydroguaiaretic acid) strongly inhibit the lipoxin formation. The complete stereochemistry of the lipoxin B formed from 15S-HETE methyl ester has been established by co-chromatography with authentic standards on various types of HPLC columns, by GC/MS analysis, by gas liquid chromatography of the ozonolysis fragments of the menthoxy carbonyl derivatives and ^{1}H-NMR studies. The molar absorption coefficient of the conjugated tetraenes was measured as $\epsilon_{301} = 53,000$.

The lipoxins formed from 15-HETE and various DiHETE isomers are formed exclusively via the oxygenation pathway as shown by experiments under an $^{17}O_2$ atmosphere and/or by anaerobic incubations. Our results indicate that lipoxins can be synthesized via lipoxygenase-catalyzed sequential oxygenation of polyenoic fatty acids and their hydro(pero)xy derivatives.

INTRODUCTION

Recently, we demonstrated that the pure lipoxygenase of rabbit reticulocytes converts 15S-HETE, 5S,15S-DiHETE and the corresponding methyl esters to a single lipoxin B isomer[1]. This product was identified as 5S,14R,15S-trihydroxy-6,8,10,12(E,Z,E,E)-eicosatetraenoate[2]. We also have direct evidence that the mechanism of this lipoxin synthesis is distinct from the pathway known to occur in human leukocytes. The pathway in human leukocytes involves the synthesis and subsequent hydrolysis of epoxide intermediates[3,4]. In contrast, the reticulocyte enzyme affects the synthesis via a series of oxygenation reactions[2].

In this report we compare and contrast the synthesis of lipoxins via oxygenation reactions catalyzed by the pure lipoxygenase of the rabbit reticulocyte and by the soybean lipoxygenase. We also include a summary of the structural elucidation of the major product of the reticulocyte reaction, and of experiments used to establish the mechanism of biosynthesis.

MATERIAL AND METHODS

<u>Materials</u>. Authentic standards of 5S,14R,15S-8<u>cis</u>-LXB, 5S,14S,15S-<u>all trans</u>-LXB and 5S,14S,15S-8<u>cis</u>-LXB were a kind gift of Drs. B.J. Fitzsimmons and J. Rokach of Merck Frosst (Canada). Other materials were as described in reference 2.

<u>Reaction conditions</u>.   Purified reticulocyte lipoxygenase[5] and soybean lipoxygenase (Sigma, type V) were incubated with the different substrates (25 µM) in 0.1 M phosphate buffer, pH 7.4 containing 10 µM 13S-hydroperoxy-9,11(Z,E)-octadecadienoic acid at 2°C. The reaction was followed recording repeated spectra of the incubation mixture in the range of 200 to 400 nm. When the absorbance at 300 nm showed no further increase, an excess of sodium borohydride was added and the products were subsequently extracted with two volumes of methylene chloride. Experiments under an atmosphere of $^{17}O_2$ were carried out as described[2].

<u>Quantitation by UV.</u> The following molar absorbance coefficients were used for quantification: $\epsilon_{235}=28,000$ (M cm)$^{-1}$ for conjugated dienes[7], $\epsilon_{242}=33,000$ (M cm)$^{-1}$ for 5,15-DiHETE[6], $\epsilon_{268}=40,000$ (M cm)$^{-1}$ for conjugated trienes[6], and $\epsilon_{300}=50,000$ (M cm)$^{-1}$ for conjugated tetraenes[8].

RESULTS

Lipoxin Formation by the Reticulocyte Lipoxygenase

The pure lipoxygenase from reticulocytes converts a variety of mono- and dihydroxy fatty acids to products containing a conjugated tetraene system. The appearance of the characteristic UV chromophore is monitored by repetitive scanning or measurement of the increase in absorbance at 300 nm. The reactions with hydroxy fatty acids require a catalytic amount of hydroperoxy fatty acid, and no spectral changes are observed in the absence of this hydroperoxide activator. Anaerobiosis and inactivation of the enzymes by 5,8,11,14-eicosatetraynoic acid also completely inhibit the reaction.

<u>Methyl esters as substrates.</u>   The rate of reaction of the reticulocyte lipoxygenase with various substrates is summarized in Table 1. Notably, the methyl esters are better substrates than are the corresponding free acids.

Table 1.   Conversion of different substrates to trihydroxy tetraenes by the pure rabbit reticulocyte lipoxygenase

12.4 µg of enzyme was incubated in 0.1 M phosphate buffer, pH 7.4 with 25 µM substrate in the presence of 5 µM 13-hydroperoxy linoleic acid at 2°C. The increase in absorbance at 300 nm during the first 3 min. of the reaction was used for the calculation of reaction rate.

substrate	reaction rate (µM/min)
15S-HETE	$\leq$ 0.01
15S-HETE methyl ester	0.11
5R/S-HETE	0.05
5R/S-HETE methyl ester	0.46

15R/15S-HETE as substrate. The influence of the chirality of the substrate is clearly seen by comparison of the reactions of 15S-HETE and 15R/S-HETE methyl esters. The pattern of products formed from these two substrates are shown in Figure 1.

The 15S-HETE methyl ester (trace A) is specifically converted to one major product absorbing at 300 nm. This compound co-chromatographed with an authentic standard of 5S,14R,15S-trihydroxy-6,8,10,12(E,Z,E,E)-eicosatetraenoic acid methyl ester (lipoxin B methyl ester) on RP-HPLC, SP-HPLC, HPLC on a silver loaded cation exchange column and on a chiral phase (dinitrobenzoyl phenyl glycine) column. The 15R/S-HETE methyl ester (trace B) is converted to a more complex mixture of products. Three products containing the characteristic tetraene chromophore were detected with LXB being the major compound. Experiments in which tracer amounts of radioactively labeled 15S-HETE methyl ester were added to the incubation mixture showed that the specific radioactivity of the products II and III were lower than that of product I. This results indicates that both products II and III are mainly formed from 15R-HETE methyl ester.

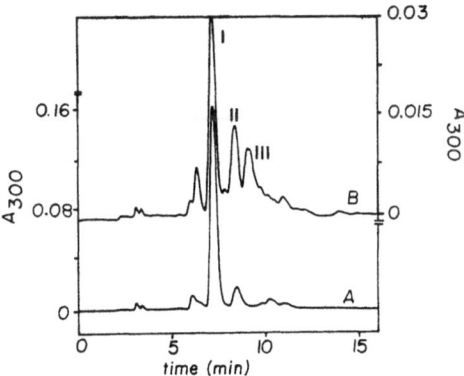

Fig. 1.   RP-HPLC analysis of the trihydroxytetraenes formed from 15S-and 15R/S-HETE methyl ester by the reticulocyte lipoxygenase.

12.4 µg/ml of reticulocyte lipoxygenase were incubated with 25 µM 15S(R/S)-HETE methyl ester in the presence of 10 µM hydroperoxy linoleic acid at 2°C. After 45 min the reaction was stopped by addition of an excess of sodium borohydride. The reaction products were prepared as described and analyzed by RP-HPLC with a solvent methanol/water/acetic acid (70/30/0.1), and a flow rate 1 ml/min. The absorbance at 300 nm was recorded. A) 15S-HETE methyl ester as substrate (left y-axis), B) 15S/R-HETE methyl ester as substrate (right y-axis).

Reactions with the natural lipoxygenase products. Figure 2 summarizes the formation of trihydroxy tetraenes from different substrates. In addition to 15S-HETE methyl ester (Fig. 1) 5S,15S-DiHETE methyl ester (A), 15S-HPETE methyl ester (B) and 14R,15S-DiHETE methyl ester are converted to LXB with a high stereospecificity. The high specificity of the conversion of 15S-HPETE methyl ester is of particular interest. In contrast to 15S-HETE methyl ester, 15S-HPETE methyl ester can be converted before or after the oxygenation at C-5 to the 14,15 epoxide derivative. Nonenzymatic hydrolysis of this compound would lead to a complex product mixture. In fact, we observed the predominant formation of LXB indicating that 15S-HPETE methyl ester is mainly metabolized via the oxygenation under our experimental conditions.

41

<u>Reactions with other substrates.</u> 14S,15S-<u>threo</u>-DiHETE methyl ester (panel D) is converted to lipoxins at a much slower rate than its <u>erythro</u> analog (0.29 versus 1.90 μmoles/ml/min at room temperature). Also, there is a more complex pattern of products. The main conjugated tetraene (50% of total) co-chromatographed with an authentic standard of 5S,14S,15S-trihydroxy-6,8,10,12(E,Z,E,E)-eicosatetraenoate, and another (40% of the total) is eluted a little earlier than 5S,14R,15S-8(Z)-LXB and corresponds in retention time to <u>all trans</u>-LXB. A product containing a conjugated diene chromophore (11.5 min, not shown) was formed in similar amounts to the tetraenes, and was only detected with the <u>threo</u> diol as substrate. This compound might be formed from a 5-hydroperoxy tetraene by a secondary reaction of the hydroperoxide; this would give a 5,6-epoxy-9,14,15-trihydroxy-eicosa-7,10,12-trienoate. 5R/S-HETE methyl ester is also converted to a complex product mixture, panel E. At least four tetraenes are formed. Each peak in the chromatogram has a broad appearance, indicating that the compounds are not homogeneous.

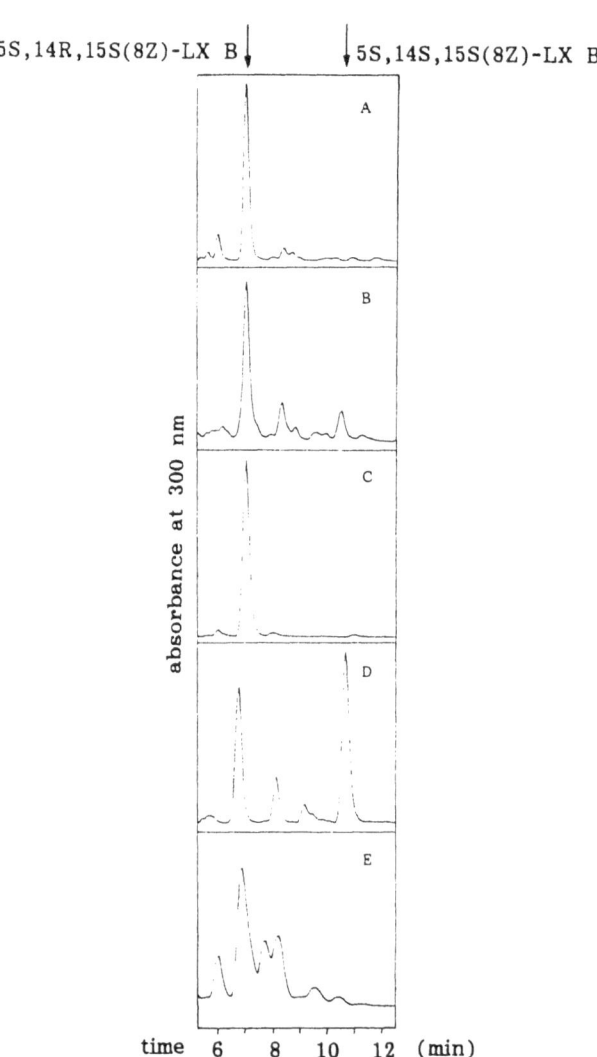

Fig. 2.   RP-HPLC analysis of trihydroxy tetraenes formed from various substrates by the reticulocyte lipoxygenase.

A) 5S,15S-DiHETE methyl ester, B) 15S-HPETE methyl ester, C) 14R,15S-DiHETE methyl, D) 14S,15S-DiHETE methyl ester, E) 5S/R-HETE methyl ester.

Lipoxin formation by the soybean lipoxygenase

This enzyme converts 15S-HPETE methyl ester to trihydroxy tetraenes with a comparable rate (Table 2) and specificity (not shown) as the reticulocyte enzyme.

In contrast to this reaction with the hydroperoxy substrate, the soybean enzyme is much less effective in forming trihydroxy tetraenes from the hydroxy fatty acids. As shown in Table 2, only small amounts of conjugated tetraenes were formed from 15S-HETE methyl ester and 5S,15S,DiHETE methyl ester. It should be stressed that the experiments with both enzymes were carried out under the same experimental conditions. The amount of enzyme added was normalized to comparable linoleic acid oxygenase activities, and the same amount of hydroperoxy activator was used with both enzymes.

Table 2. Relative rates of formation of conjugated tetraenes by the lipoxygenases from soybeans and reticulocytes

Substrate	Relative reaction rate:	
	reticulocyte lipoxygenase	soybean lipoxygenase
15S-HPETE methyl ester	1	0.7
15S-HETE methyl ester	1	<0.1
5S,15S-DiHETE methyl ester	1	0.2

The soybean and reticulocyte enzymes show different reaction kinetics. With 5S,15S-DiHETE methyl ester as substrate, the soybean lipoxygenase forms conjugated tetraenes only within the first 10 minutes after enzyme addition, whereas the reaction with the reticulocyte enzyme continues for 3 hours. This short term reaction is not due to self inactivation of the enzyme - this was tested by measurement of linoleic acid oxygenase activity after the formation of tetraenes had stopped. It is likely that the hydroperoxy activator is more rapidly consumed in the reaction of the soybean enzyme with 5S,15S-DiHETE methyl ester.

Structural elucidation of the major product of reaction of 15S-HETE methyl ester with reticulocyte lipoxygenase (Compound I, Fig 2)

The UV-spectrum shows a conjugated tetraene chromophore with maxima of strong absorbance at 287, 301, and 315 nm (Figure 3). The mass spectrum (Fig. 4) with the characteristic ions low intensity at m/z 582 (M+), 492 (M+-90), 482 (M+-100), and 402 (M+-180) indicate an eicosanoid with three silylated OH-groups and four double bonds. The positions of the double bonds are indicated by the ions at m/z 203, 275, 173, and 409 as shown in the insert in Fig. 4. The mass spectrum of the hydrogenated compound (not shown) is in line with these results.

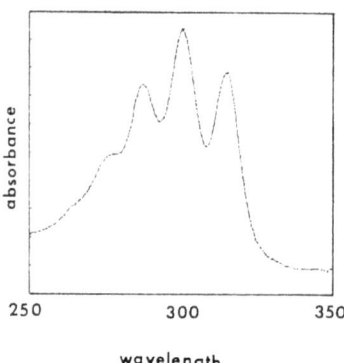

Fig. 3.   UV spectrum of the major product after the oxygenation of 15S-HETE methyl ester by the reticulocyte lipoxygenase (Compound I)

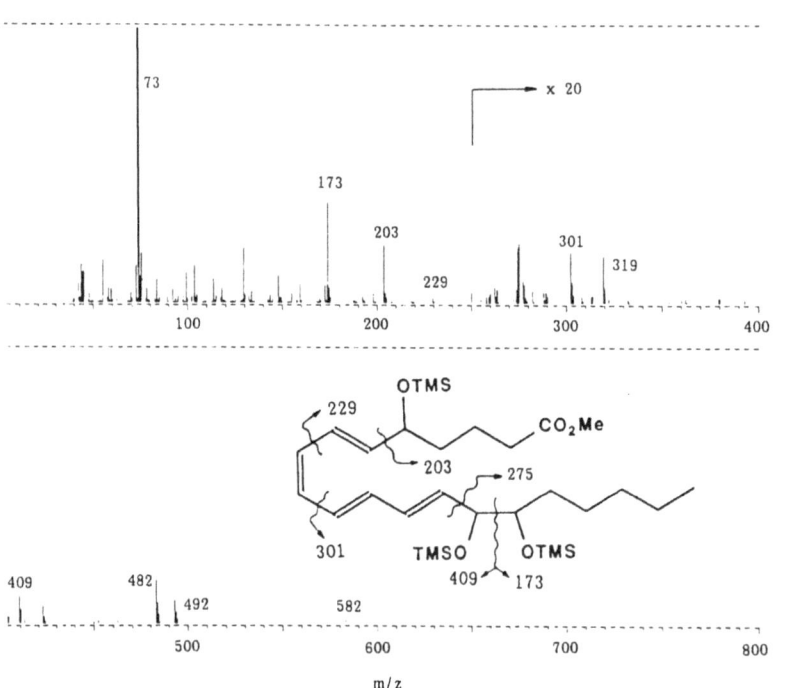

Fig. 4.   GC/MS analysis of compound I formed from 15S-HETE methyl ester

In order to establish the stereochemistry at carbon-5, compound I was converted to its trimenthoxy carbonyl derivative and subjected to oxidative ozonolysis. After methylation, the resulting 2OH-adipic acid derivatives were analyzed by gas/liquid chromatography. As shown in **Fig. 5**, more than 90% of the compound I derivative co-chromatographed with the authentic standard of 2-OH (L)-adipic acid. This assignment corresponds to the D_S-configuration at C_5 in the trihydroxy tetraene. The relative configuration of the 14,15 diol is erythro (established by comparison with authentic erythro and threo lipoxin standards). Since the absolute configuration at C_{15} is S (the same as in the substrate), the absolute configuration at C_{14} has to be R.

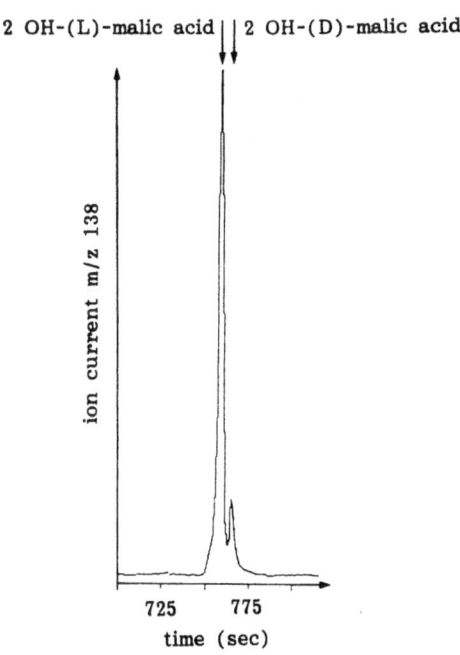

2 OH-(L)-malic acid | | 2 OH-(D)-malic acid

Fig. 5.   Absolute configuration at C-5. Gas liquid chromatography of the ozonolysis fragments of the menthoxy carbonyl derivative, ozonized, methylated with diazomethane and analyzed by gas chromatography on a fused silica column SPB-1 (30 m x 0.25 mm, coating thickness 0.25 µm), flow 2 ml/min. Temperature program: 70°C - 230°C with 20°C/min, 230°C - 280°C with 2°C/min. The Varian Vista 6000 gas chromatograph was coupled to a Nermag 10-10C mass spectrometer. The ion current at m/z 138 was recorded[10].

Figure 6 shows the complete ^{1}H-NMR spectrum of compound I (free acid). The two dimensional correlation spectroscopy (COSY spectrum) allows the assignment of all protons of the molecule. The coupling constants over the double bonds $J_{6,7}=15.8$ Hz, $J_{8,9}=8.4$, $J_{10,11}=14.5$ Hz and $J_{12,13}=14.6$ Hz indicate a 6E,8Z,10E,12E conjugated system.

These analyses prove that compound I formed by the pure lipoxygenase is the methyl ester of 5S,14R,15S-trihydroxy-6,8,10,12(E,Z,E,E)-eicosatetraenoate. Its chemical structure is identical with that of the lipoxin B formed by human leukocytes[3].

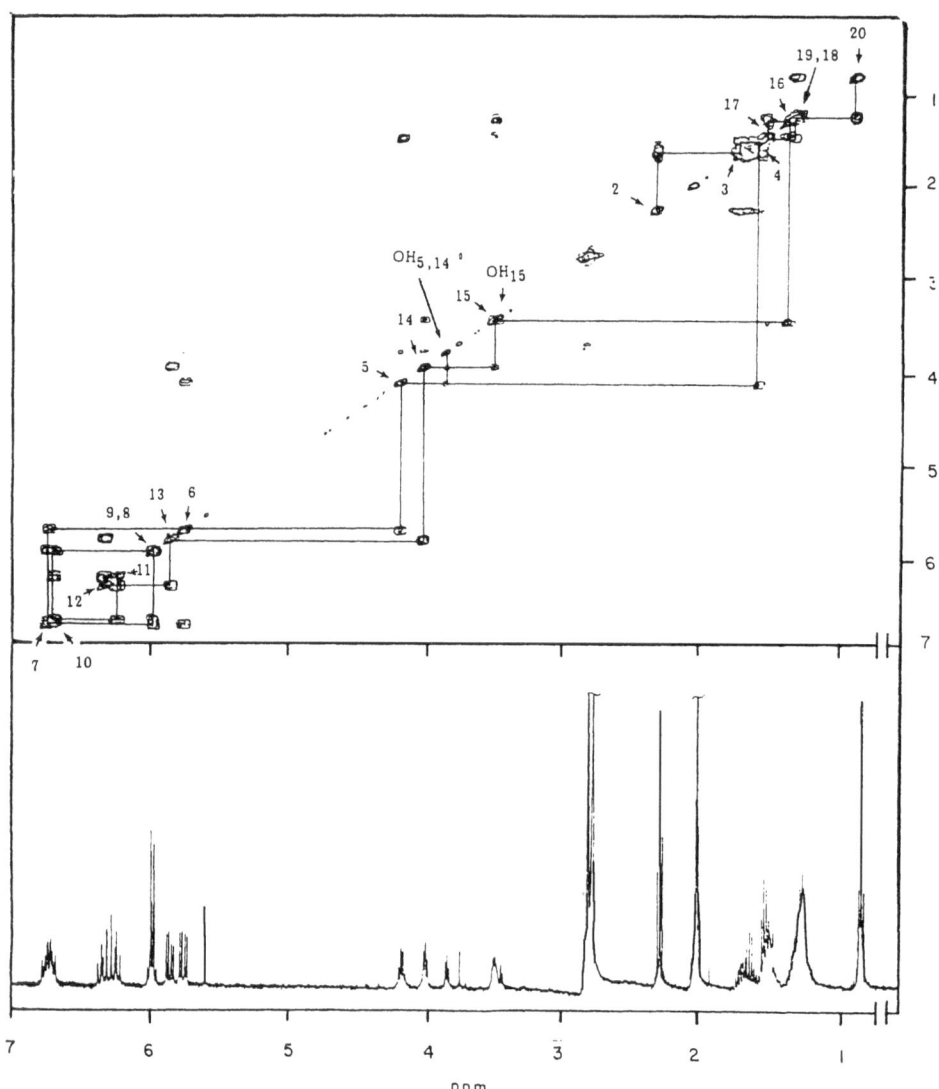

Fig. 6. ^{1}H-NMR spectrum (400 MHz) of compound I (free acid)

Compound I (800 µg) was hydrolyzed to the free acid, ca.10% all-trans impurity was removed by RP-HPLC, and the spectrum was recorded in deuterated acetone with a Brucker AM 400 instrument at 20°C.

Estimation of the molar absorption coefficient of LXB

The substrate was [5,6,8,9,11,12,14,15-^{3}H$_8$]15S-HETE methyl ester. None of the labeled hydrogens is directly involved in the oxygenation reactions, and therefore a quantitatively major isotope effect is not expected. Therefore we made the assumption that the specific radioactivity of the product is the same as in the substrate.

Figure 7 shows RP-HPLC purification of the radiolabeled LXB. Fractions of 12 seconds were taken for counting the radioactivity. The ^{3}H-labeled LXB was found to elute about 0.5 min earlier than the unlabeled compound. The fractions

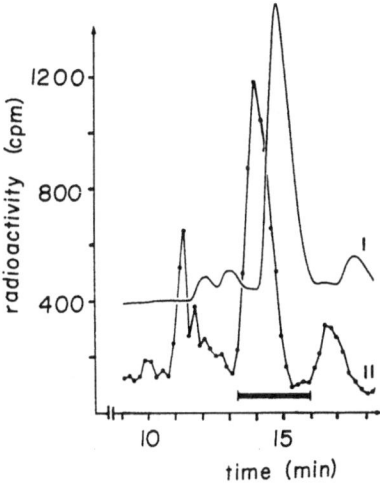

Fig. 7.   RP-HPLC analysis of compound I formed from ^{3}H-labeled 15S-HETE methyl ester.

12.4 µg/ml reticulocyte lipoxygenase was incubated with 25 µM 5,6,8,9,11,12,14,15-octatritiated 15-HETE methyl ester (2.1 Ci/mol) for 45 min at 2°C. RP-HPLC as described in legend to Fig. 2. Fractions of 12s were collected and aliquots were counted for radioactivity. The fractions indicated below the traces were pooled and used for the determination of the molar absorption coefficient. I: absorbance at 300 nm; II: radioactivity.

were pooled as shown in the figure, and checked for purity by RP-HPLC analysis of an aliquot. In the main sample, measurement of the radioactivity and the absorbance at 301 nm gave a molar absorption coefficient $\epsilon_{301} = 53,000$ (M cm)$^{-1}$. This value is little higher than that reported earlier for conjugated tetraenes (50,000, ref. 8).

The molar absorption coefficient for the minor product, 5,15-DiHETE methyl ester, was measured in the same experiment. The value of $\epsilon_{242} = 31,000$ (M cm)$^{-1}$ is close to the value of 33,000 reported earlier[6].

Mechanism of formation of Lipoxin B

The main question to be answered was whether LXB is formed from 15S-HETE methyl ester via two sequential oxygenation steps or via hydrolysis of an epoxide intermediate? The 15-hydroxy substrate cannot be converted to a 14,15-epoxide, but there is a possibility for formation of a 5,6-epoxide. Indirect evidence against this mechanism is the finding of a single major product from most of the substrates, as opposed to the more complex pattern expected via a nonenzymatic epoxide hydrolysis.

For direct proof of the oxygenase pathway, the main lipoxin product was prepared from 15S-HETE methyl ester under an atmosphere of $^{17}O_2$. GC/MS analysis of the LXB indicates the incorporation of two atoms of $^{17}O_2$. As shown in **Table 3**, the ions containing the OH-group at C_5 (m/z 204 and 230) are shifted by one mass unit as compared with the corresponding ions of the product formed under and $^{16}O_2$-atmosphere. The ions at m/z 302 and 276 containing the OH-group at C_{14} are also shifted by one mass unit. The ions at m/z 411, 484, and 494 containing both OH-groups are shifted by two mass units. The percentage incorporation of ^{17}O almost exactly matched the content of the ^{17}O-labeled oxygen gas[2].

Table 3. GC/MS analysis of the lipoxin B formed from 15S-HETE methyl ester under an atmosphere of $^{16}O_2$ or $^{17}O_2$.

ion	$^{16}O_2$-LX B	$^{17}O_2$-LX B
$C_6H_{12}OTMS$	173 (34.2)	173 (43.1)
$C_5H_7(OTMS)O_2Me$	203 (20.7)	204 (20.5)
$C_7H_9(OTMS)O_2Me$	229 (2.4)	230 (1.9)
$C_{11}H_{17}(OTMS)_2$	301 (1.1)	302 (0.9)
$C_7H_{13}(OTMS)_2$	275 (1.3)	276 (1.4)
$M^+-(C_6H_{12}OTMS)$	409 (0.5)	411 (0.5)
M^+-100	482 (0.7)	484 (0.5)
M^+-90	492 (0.4)	494 (0.3)
M^+	582 (0.1)	

Thus, it could be concluded that the LXB was formed exclusively via the oxygenation pathway. Kinetic studies showed that 5S,15S-DiHETE methyl ester is an intermediate in this reaction[2].

DISCUSSION

Methyl esters are the preferred substrates: The pure lipoxygenases from reticulocytes and soybeans convert a variety of mono- and dihydroxy fatty acids to products containing a conjugated tetraene system. It is interesting that the methyl esters of the hydroxy fatty acids are better substrates that the free acids. This difference is most prominent with 15S-HETE. Whereas the free acid is mostly converted to various DiHETEs with LXB being only a minor product, its methyl ester is mainly converted to LXB. The possible reasons for this behavior have been discussed before[2]. Assuming there is only one active site on the enzyme for the catalysis of stereospecific oxygenations, then the fatty acid substrate must be oriented the opposite way round for 15S and 5S oxygenations. The experimental observation is that the methyl ester is more efficiently converted to the 5S,15S-DiHETE intermediate. Thus, the increased hydrophobicity of the methyl ester may favor this inverse orientation at the active site of the enzyme, and result in a far more efficient conversion to the 5,15-DiHETE. This effect increases the reaction rate by an order of magnitude.

The methyl esters are also better converted from the 5,15-DiHETE intermediate to the LXB product. This effect is seen with both the soybean and the reticulocyte enzymes, and contributes a 3-4 fold increase in reaction rate compared with the free acid substrate.

Influence of steric configuration of the substrate: Hydroxy fatty acids which are formed from arachidonate by the reticulocyte enzyme (15S-H(P)ETE, 5S,15S-DiHETE, 14R,15S-DiHETE) are good substrates for the tetraene formation in terms of a high reaction rate and/or high product specificity. The "unnatural" products 5R/S-HETE, 14S,15S-DiHETE and 15R-HETE are converted to tetraenes with a lower rate and a lower specificity. 15RS-HETE reacts at a slower rate than the same concentration of 15S-HETE, and a more complex pattern of products are formed from both enantiomers. The first oxygenation step, the formation of 5S,15R/S-DiHETE methyl ester, is not influenced by the configuration at C_{15}, since comparable amounts of 15R- and 15S-HETE methyl esters were recovered. Therefore, one has to conclude that mainly the second oxygenation step converting 5S,15(R/S)-DiH(P)ETE methyl ester is influenced by the steric configuration of C_{15}.

Comparison of reticulocyte and soybean lipoxygenase: As shown in **Table 2**, the reticulocyte lipoxygenase is more effective in the formation of trihydroxy tetraenes from all substrates tested. However, it should be stressed that these data were obtain in an assay optimized for the reticulocyte enzyme. The differences in the kinetics of tetraene formation from 5S,15S-DiHETE (see results) suggest that the low formation of tetraenes by the soybean enzyme is partly due to kinetic properties of the reaction. It is known that this enzyme has a low affinity (K_M-value) and a low reaction rate (V_{max}) for the oxygenation of 15S-HPETE[6]. In contrast, the affinity of the reticulocyte enzyme for 15S-HETE is comparable with that for arachidonic acid (unpublished). Furthermore, the concentration of 13S-hydroperoxy linoleic acid necessary as activator for the enzyme is of importance for the reaction rate. For the reticulocyte enzyme it has been shown that a certain activator/substrate ratio is necessary for the maximal oxygenation of 15S-HETE[9]. This behaviour suggests that there is a competition between the activator and the substrate for the binding site(s) at the enzyme. The soybean enzyme may require higher concentrations of activator for its maximal activity. The relatively high LXB formation with 15S-HPETE as substrate in an assay in which the substrate itself acts as activator seems to support this hypothesis.

ACKNOWLEDGEMENTS

We thank Dr. Thomas M. Harris and Steven W. Baertschi for recording the ^{1}H-NMR spectrum. This work was supported by NIH grants GM15431 and DK35275.

REFERENCES

1. Kühn, H., Wiesner, R., Alder, L., Schewe, T. and Stender, H. FEBS Lett. 208, 148-152 (1986).
2. Kühn, H., Wiesner, R., Alder, L, Fitzsimmons, B.J., Rokach, J., and Brash, A.R. Eur. J. Biochem., in press (1987).
3. Serhan, C.N., Hamberg, M., and Samuelsson, B. Proc. Natl. Acad. Sci. USA 83, 1983-1987 (1986).
4. Fitzsimmons, B.J., Adams, Evans, J.F., LeBlanc, Y., and Rokach, J. J. Biol Chem. 260, 13008-13012 (1985).
5. Schewe, T., Wiesner, R. and Rapoport, S.M. Meth. Enzymol. 71, 430-441 (1981).
6. van Os, C.P.A., Rijke-Schilders, G.P.M., van Halbeek, H., Verhagen, J. and Vliegenthart, J.F.G. Biochim. Biophys. Acta 663, 177-193 (1981).
7. Tappel, A.L., Boyer, P.D. and Lundberg, W.D. J. Biol. Chem. 199, 267-281 (1952).
8. Serhan, C.N., Hamberg, M., and Samuelsson, B. Proc. Natl. Acad. Sci. USA 81:5335-5339 (1984).
9. Kühn, H., Wiesner, R., Stender, H., Schewe, T., Lankin, V.Z., Nekrassov, A. and Rapoport, S.M. FEBS Lett. 203:247-252 (1986).
10. Hammarstrom, S. Meth. Enzymol. 35:326-334 (1975).

5

BIOSYNTHESIS AND BIOLOGICAL ACTIVITIES OF LIPOXIN A_5 AND B_5 FROM

EICOSAPENTAENOIC ACID

Bing K. Lam and Patrick Y-K Wong

Department of Pharmacology
New York Medical College
Valhalla, New York 10595

SUMMARY

[1-^{14}C]-Eicosapentaenoic acid (EPA) was incubated with porcine leukocytes. Three polar metabolites were isolated after RP-HPLC separation in addition to pentaene leukotrienes and mono-hydroxy fatty acids. These compounds display U.V. absorbance with U.V. λmax at 302 nm with shoulders at 289 and 317 nm which were typical of a conjugated tetraenes. Using an alkaline RP-HPLC solvent system, it was found that these three compounds co-eluted with synthetic standards of lipoxin A_5, lipoxin B_5 and 5S,6S,15S-lipoxin A_5 [6-S-LXA$_5$]and were identified accordingly. Their structures were further confirmed by GC/MS analysis. When tested for biological activities, it was found that both lipoxin A_5 and lipoxin B_5 induce superoxide anion generation in canine neutrophils. Furthermore, LXA$_5$ caused a dose-related contraction of isolated rat tail artery. The biological potency of 5-series lipoxins were similar to those of 4-series.

INTRODUCTION

Lipoxins, the new series of biologically active derivative of arachidonic acid, were first isolated by Serhan and coworker[1,2,3] after incubating 15-hydroperoxyeicosatetraenoic acid (15-HPETE) with human leukocytes. Their structures were elucidated as 5S,6R,15S-trihydroxy-7,9,13-E,11-Z-eicosatetraenoic acid or lipoxin A_4 (LXA$_4$) and 5S,14R,15S,trihydorxy-6,10,12-E-8-Z-eicosatetraenoic acid or lipoxin B_4 (LXB$_4$) as well as their structural isomers[4,5]. These newly discovered lipoxins display a characteristic U.V. absorbance with λmax at 301 nm and shoulders at 287 and 316 nm due to the presence of conjugated tetraenoic structure[6]. Furthermore, these compounds possess

Abbreviations: LXA$_5$, 5S,6R,15S-trihydroxy-7,9,13-E-11,17-Z-eicosapentaenoic acid; LXA$_4$, 5S,6R,15S-trihydroxy-7,9,13-E-11-Z-eicosatetraenoic acid; LXB$_5$, 5S,14R,15S-trihydroxy-6,10,12-E-8,17-Z-eicosapentaenoic acid; LXB$_4$, 5S,14R,15S-trihydroxy-6,10,12-E-8-Z-eicosatetraenoic acid; 15-HPETE, 15-hydroperoxyeicosapentaenoic acid; 15-HPETE, 15-hydorperoxyeicosatetraenoic acid; 15-HEPE, 15-hydroxyeicosapentaenoic acid; 15-HETE, 15-hydroxyeicosatetraenoic acid; EPA, eicosapentaenoic acid; AA, arachidonic acid; PBS, phosphate buffered saline.

potent biological activities on variety of assays system. LXA$_4$ contracted Guinea-pig lung parenchymal strips[7]; induces superoxide anion generation in human neutrophils[2]; leukocyte chemokinesis[8]; activation of protein kinase C[9], and both LXA$_4$ and LXB$_4$ inhibited human natural killer cytotoxicity[10]. The distinct biological activities as well as the recent finding by our laboratory the lipoxin can be synthesized endogenously by porcine leukocyte[11] ensure the further investigation of this new series of eicosanoids.

15-hydroperoxyeicosapentaenoic acid (15-HPEPE), an EPA derived product has previously been shown to be converted to LXA$_5$ and LXB5 by porcine leukocytes by our laboratory[3]. In the present study, we examine the biotransformation of EPA into lipoxin A$_5$ and B$_5$ in porcine leukocytes.

MATERIALS AND METHODS

Materials

[1-^{14}C] EPA (Specific activity, 55 mCi/mmole) was purchased from New England Nuclear, Boston, MA. 5,8,11,14,17-eicosapentaenoic acid was obtained by treatment of eicosapentaenoic acid ethylester (99.7% pure, a gift from Central Research Laboratory, Nippon Suisan Kaisha, Tokyo, Japan) with 40% methanolic KOH overnight at room temperature. Calcium ionophore A23187 was from Calbiochem-Behring (San Diego, CA). DiHEPEs, LTB$_5$ and mono-HEPEs were gifts from W.W. Diagnostic Products (West Haverstraw, N.Y.). LXA$_5$, LXB$_5$ and 6-S-LXA$_5$ standards were gifts from Dr. Spur. Dextran T-500 was from Pharmacia, Sweden.

Preparation of Porcine Leukocytes

Porcine leukocytes were prepared from mixed venous blood containing 10 mM EDTA[12]. Venous blood was mixed with half volume of 6% dextran solution and allowed to settle for 60 min at room temperature. The upper phase was then centrifuged at 250 x g for 20 min and the contaminating red cells were removed by hypotonic lysis with distilled water for 20 sec. The final cells suspension was reconstituted by the addition of an appropriate volme of 3.6% saline. After centrifugation at 250 x g for 20 min, the pellet was resuspended in Dulbecco's phosphate-buffered saline (pH 7.4) to a final concentration of 100x10^6 cells/ml. The viability of the cells as measured by Trypan blue exclusion test was found to be greater than 95%.

Incubation, Extraction and Purification

After prewarming the leukocyte suspension to 37°C for 5 min, [1-^{14}C]-EPA and ionophore A23187 (in ethanol) were added to final concentration of 100 μM and 5 μM, recpectively, and incubated for 30 min at 37°C with constant shaking. The amount of ethanol needed to dissolve EPA and ionophore never exceeded 0.1% of the incubation volume. The incubation was terminated by the addition of 2 vol ethanol. The incubation precipitate was filtered and the ethanolic filtrate was evaporated to near dryness. the residue was dissolved in 5 ml of distilled water and acidified to pH 4.0 with 1 M HCl. The solution was then extracted with 10 vol. of ethyl acetate. The ethyl acetate fraction was evaporated to dryness under N$_2$. The residue was dissolved in 50 μl of methanol and separated by HPLC on a Water's Associates Dual Pump system equipped with a reverse-phase ultrasphere ODS colum

52

(C_{18}-ODS, 5μ, 10 mm x 25 cm, Beckman, Palo Alto, CA), a U-6K injector and a 481 variable wavelength detector. The products were eluted with a linear gradient from methanol/water/acetic acid (50:50:0.05, v/v) (solvent A) to methanol (solvent B) for 40 min at a flow rate of 3 ml/min[11]. Column effluents were monitored with a Water's Associates 481 λmax variable wavelength detector set at 302 nm (0-8 min), 270 (9-20 min) and 237 nm (21-40 min) for porcine leukocytes. Fraction of 3 ml were simultaneously collected with an on-line fraction collector and a 50 μl portion of each fraction was removed for estimation of recovered radioactivity.

Ultraviolet Spectroscopy

Samples eluted from the HPLC were evaporated to dryness under vacuum, dissolved in absolute ethanol and examined with a Hewlett-Packard 8450-A ultraviolet/visible spectrophotometer.

Gas Chromatography-Mass Spectrometry

The methylester of each sample was converted to trimethylsilyether derivatives by additon of 25 μl of pyridine followed by 50 μl of trimethylchlorosilane and 50 μl of hexamethyldisilazane (Supelco). The mixtures were kept at room temperature for 20 min and evaporated to dryness with N_2. Next, the samples were dissolved in hexane (25 μl) and injected into the gas chromatograph-mass spectrometer (Hewlett-Packard 5895-B) equipped with cross-linked methylsilicone capillary column (12 mm x 0.30 cm). The helium flow was set at 16 ml/min, with the oven temperature, injection temperature and ion source temperature set at 200, 260 and 200°C, respectively. The electron energy was set at 70 eV.

RESULTS

RP-HPLC separation of the sample extracts from incubation of [1-^{14}C]-EPA with porcine leukocytes revealed a polar fraction (fraction X) with highest UV absorbance maxima at 301 nm which eluted very close to the solvent front (Fig. I). When examined by U.V. spectrometry, it showed U.V. absorbance with U.V. λmax at 301 nm and shoulders at 289 and 317 nm indicates that this fraction contains lipoxin-like material. The formation of Fraction X by porcine leukocytes was time-dependent which plateaus after 15 min of incubation (Fig. 2). In order to determine the nature of this fraction, Fraction X isolated from different experiments were pooled and further purified by a second RP-HPLC using Hamilton polystylene column (Universal Scientific Inc., Alanta, GA) and was eluted with an alkaline solvent system consisting of 0.1 M boric buffer pH 8.8 and methanol with a ratio of 4:6. Under this condition, Fraction X was separated into 4 components (Fig. 3). Two of the components co-eluted with the synthetic standards of LXA_5 and LXB_5 (Fig. 2). To further elucidate the identities of these two compounds, materials eluted from HPLC were methylated and derivatized forGC/MS structural analysis. Mass spectrum (C-value=24.3) of the material co-eluted with authentic LXA_5 standard showed ions of high interactions at M/Z 203, 171, 129 ions of lower intensities were observed at M/Z 580, 490, 480, 377, 409 (Fig. 4A). The mass spectrum as well as C value was consistent with those reported for LXA_5[3]. When analyzed by GC/MS, mass spectra (C-value=25.4) of material co-eluted with LXB_5 showed ion of high intensities at 171, 129 and 203. Ions of lower intensities were observed at M/Z 580, 565, 490, 480, 377 (Fig. 4B). These results were again consistent with those reported for LXB_5[3].

To investigate if EPA derived lipoxins also possess biological activity as AA derived product LXA$_5$ and B$_5$ isolated from incubation extract were tested on superoxide anion generation and aggregation on canine neutrophils and musculotropic activity in isolated rat tail artery. Similar to AA derived LXA$_4$, LXA$_5$ and LXB$_5$ induced

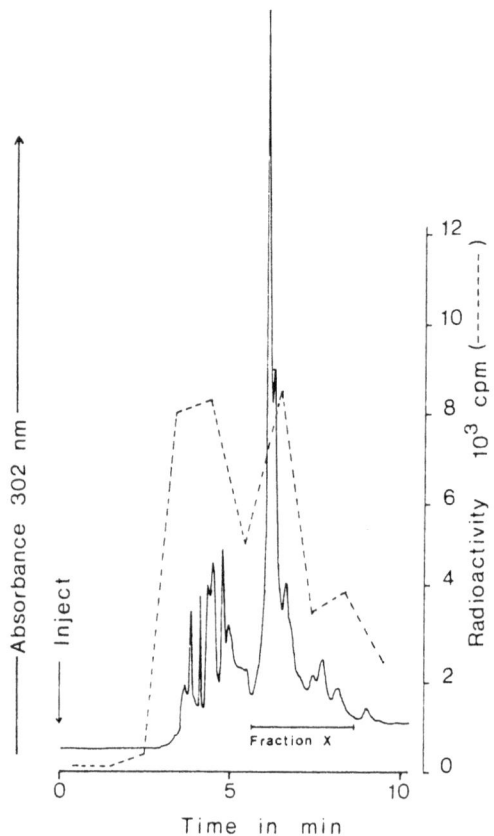

Figure 1. Partial RP-HPLC chromatograph of sample extract isolated from incubation of [1-^{14}C]-EPA with porcine leukocytes.

superoxide anion generation with similar potency to LXA$_4$ but was about 1/4 as potent as the chemotactic peptide f-Met-Leu-Phe (Fig. 5). In addition, LXA$_5$ induced contraction of isolated rat tail artery in a dose dependent manner similar to AA derived LXA$_4$ (Fig. 6).

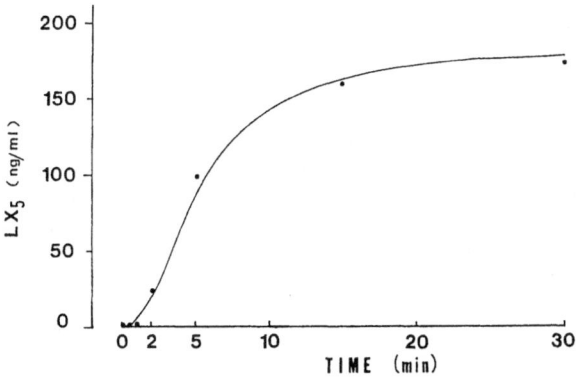

Figure 2. Formation of fraction X (lipoxin fraction) with various incubation time.

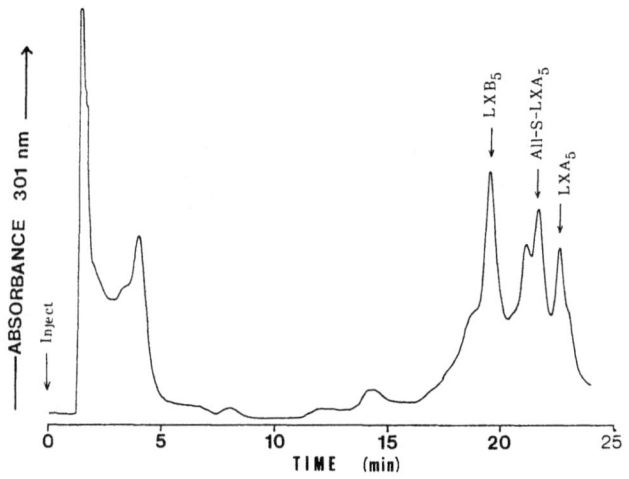

Figure 3. RP-HPLC repurification of fraction X with Hamilton polystylene column. Arrows indicate the retention time of synthetic standards of LXA_5, LXB_5 and 6-S-LXA_5 (All-S-LXA_5)

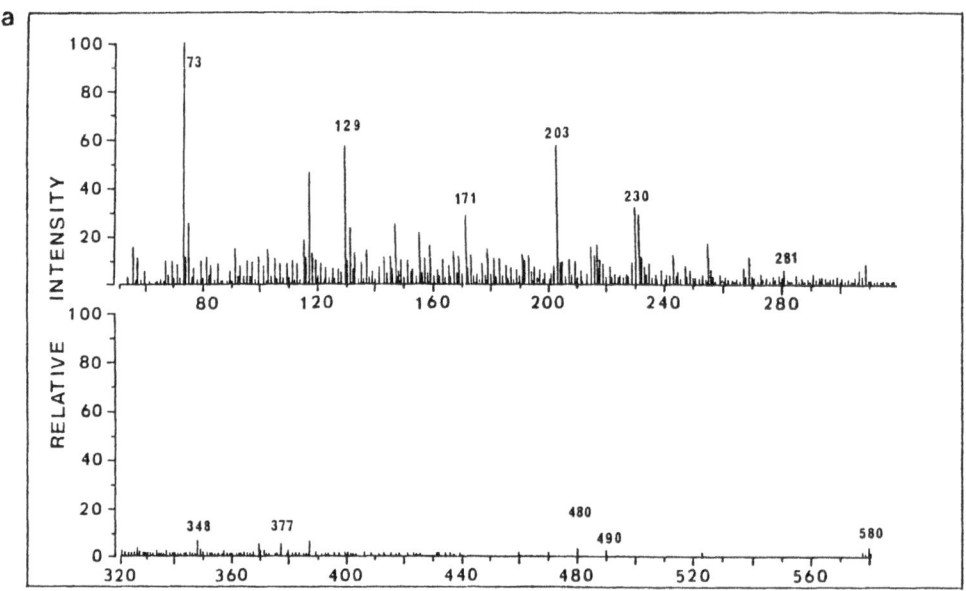

Figure 4A. Mass spectrum of ME-TMS derivative of the material coeluted
with the synthetic standards of LXA$_5$(A) and LXB$_5$(B).

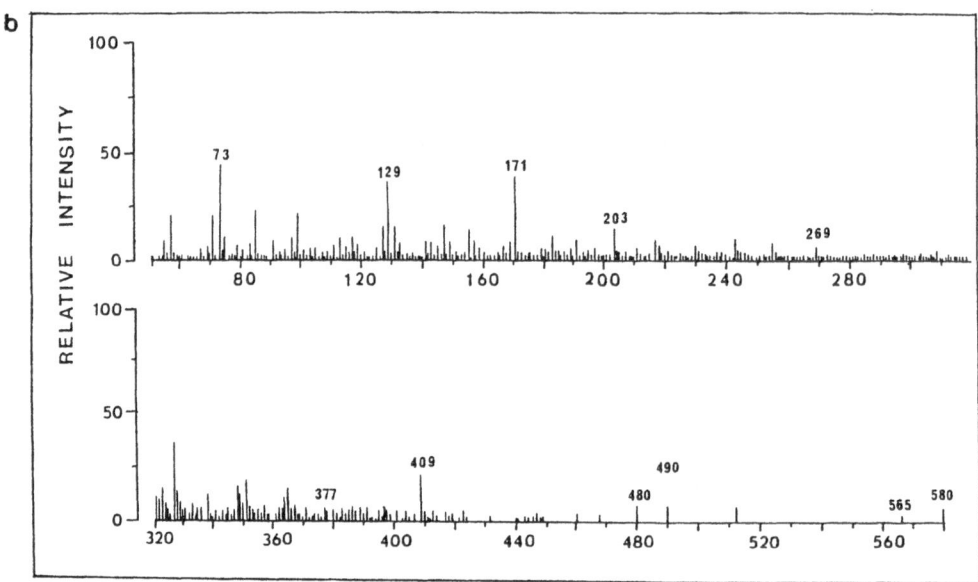

Figure 4B. Mass spectrum of ME-TMS derivatives of the material coeluted
with synthetic standard of LXB$_5$

DISCUSSION

Previously we have shown the conversion of 15-HPEPE into LXA_5 and LxB_5 by porcine leukocytes. In this study, we describe the generation of lipoxin of 5-series directly from EPA. Their structural elucidations were based on their chromatographic behavior on RP-HPLC, U.V. spectrum, C-values and by GC/MS. Although four compounds were isolated after Hamilton column RP-HPLC separation, only the materials which co-eluted

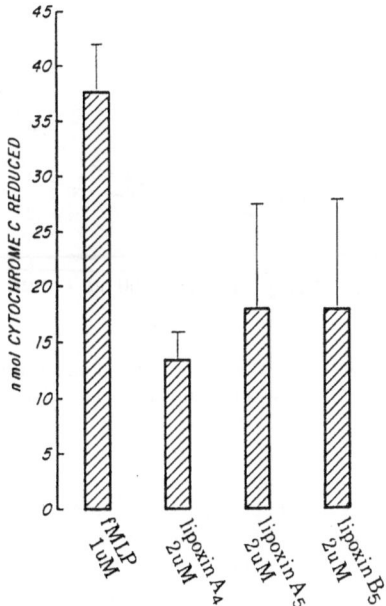

Figure 5. Superoxide anion generation induced by LXA_4, LXA_5, LXB_5 and fMLP.

with authentic standards of LXA_5 and LXB_5 were analyzed by GC/MS and were identified accordingly. $6\text{-}S\text{-}LXA_5$ was identified by co-eluting with synthetic standard of $6\text{-}S\text{-}LXA_5$, no mass spectrometry was performed on this sample. Materials eluted just after LXB_5 was not identified. In addition, its is important to note that $6\text{-}S\text{-}LXA_5$ eluted before LXA_5 in our RP-HPLC system which is opposite to those reported by Serhan et al. for LXA_4 and $6\text{-}S\text{-}LXA_4$[4].

Recently, Serhan and coworkers[4,5] and Ueda et al.[13] have reported the mechanism of lipoxin formation using 15-HPETE as substrate. The formation involves an initial oxygenation at C5 position yielding a 5,15-dihydroperoxy derivative followed by an epoxide formation at C-5 and C-6 very likely to be catalysed by LTA_4 synthetase that is associated with 5-lipoxygenase[14]. In addition Yamamoto and coworker reported that 12-lipoxygenase can catalyse the formation fo LXB_4 using 5,15-diHPETE as substrate[15]. Although there are several biosynthetic pathways for the formation of lipoxins, in this study we have not determined which pathway is involved in the formation of lipoxins from EPA.

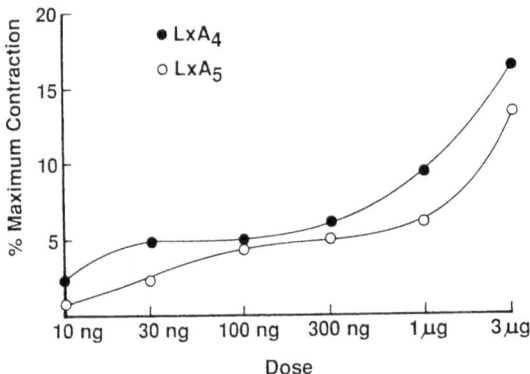

Figure 6. Dose dependent contraction of isolated rat tail artery (expressed as percent of maximum contraction induced by phenylephrine, 10-30 μg).

Arachidonic acid derived lipoxins of 4-series have been shown to possess potent biological activities which include inducing superoxide anion generation in human neutrophil activation of protein kinase C and inhibition of natural killer cells cytolytic activity[9,10]. We tested LXA_5 and LXB_5 on superoxide anion generation in canine neutrophils and compared their potency to that of LXA_4 and fMLP. Similar to LXA_4, both LXA_5 and LXB_5 induce superoxide anion generation in canine neutrophils (Fig. 6). The potency of these two compared were similar to that of LXA_4 and was one quarter as potent as fMLP (Fig. 6). Unlike superoxide anion generation, neither LXA_5, LXB_5 nor LXA_4 induces lysosomal enzyme release or causes aggregation in either canine or rat neutrophils (data not shown). Furthermore, the observation that both LXA_4 and LXA_5 contracted isolated rat tail artery suggest a possible role of these lipid mediators in cardiovascular homeostasis (Fig. 6).

In this report, we have demonstrated that porcine leukocytes can metabolize EPA to formation lipoxins of 5-series. Unlike LTB_5 which was less potent than LTB_4, these trihydroxypentaene derivatives of EPA possess biological activies comparable to their arachidonic acid-derived counterparts (Fig. 5). Previously, we demonstrated that 15-HEPE was the major metabolites of 15-HPEPE in human platelet and porcine leukocytes[16,12] and exhibited a more potent biological activity than 15-HETE on neutrophil aggregation as provoked by fMLP[16]. Thus, it is possible that during nutritional manipulation, diets enriched with EPA or fish oil may result in the generation of biologically active lipid mediators derived from platelets, leukocytes and other cell types during cell-cell interaction.

REFERENCES

1. Serhan, C.N., Hamberg, M., and Samuelsson, B. Trihydroxytetraenes: A novel series of compounds formed from arachidonic acid in human leukocytes. Biochem. Biophys. Res. Commun., 118:943-949 (1984)

2. Serhan, C.N., Hamberg, M., and Samuelsson, B. Lipoxins: A novel series of biologically active compounds formed from arachidonic acid in human leukocytes. Proc. Natl. Acad. Sci. U.S.A. 81:5335-5339 (1984)

3. Wong, P.Y-K., Hughes, R.A., and Lam, B. Lipoxene: A new group of trihydroxy pentaene of eicosapentaenoic acid derived from porcine leukocytes. Biochem. Biophys. Res. Commun. 126:763-772 (1985)

4. Serhan, C.N., Nicolaou, K.C., Webber, S.E., Veale, C.A., Dahlen, S.E., Puustinen, T.J., and Samuelsson, B. Lipoxin A: Stereochemistry and Biosynthesis. J. Biol. Chem. 261:16340-16345 (1986)

5. Serhan, C.N., Hamberg, M., Samuelsson, B., Morris, J., and Wishka, D.G. On the stereochemistry and biosynthesis of lipoxin B. Proc. Natl. Acad. Sci. U.S.A. 83:1983-1987 (1985)

6. Murphy, R.C., Hammarstrom, B., and Samuelsson, B. Leukotriene C, a slow reacting substance from murine mastocytoma cells. Proc. Natl. Acad. Sci. U.S.A. 76:4275-4279 (1979)

7. Dahlen, S.E. Biological activities of lipoxins: FASEB Symposium on "Lipoxin: Biosynthesis and Pharmacology, Washington, D.C., March 30 (1987)

8. Lee, T.K. Biological activities of lipoxins of 5-series. FASEB Symposium on "Lipoxin: Biosynthesis and Pharmacology, Washington, D.C., March 30 (1987)

9. Hansson, A., Serhan, C.N., Haeggstrom, J., Ingelman-Sundberg, M., and Samuelsson, B. Activation of protein kinase C by lipoxin A and other eicosanoids. Intracellular action of oxygenation products of arachidonic acid. Biochem. Biophys. Res. Commun. 134:1215-1222 (1986)

10. Ramstedt, U., Ng., J., Wigzell, H., Serhan, C.N., and Samuelsson, B. Action of novel eicosanoids lipoxin A and lipoxin B on human natural killer cell cytotoxicity: Effects on intracellular cAMP and target cell binding. J. Immunol. 135:3434-3438 (1985)

11. Lam, B.K., Serhan, C.N., Samuelsson, B., and Wong, P.Y-K. A phospholipase A_2 isoenzyme provokes lipoxin B formation from endogenous sources of arachidonic acid in porcine leukocytes. Biochem. Biophys. Res. Commun. 144:123-131 (1987)

12. Lam, B.K., Hirai, A., Yoshida, S., Tamura, Y., and Wong, P.Y-K. Transformation of 15-hydroperoxyeicosapentaenoic acid to lipoxin A_5 and B_5, mono- and dihydroxyeicosapentaenoic acids by porcine leukocytes. Biochim. Biophys. Acta. 917:398-405 (1987)

13. Ueda, N., Yamamoto, S., Fitzsimmons, B.J., and Rokach, J. Lipoxin synthesis by arachidonate 5-lipoxygenase purified from porcine leukocytes. Biochem. Biophys. Res. Commun. 144:996-1002 (1987)

14. Rouzer, C.A., Massumato, T., and Samuelsson, B. Single protein from human leukocytes possesses 5-lipoxygenase and leukotriene A_4 synthetase activities. Proc. Natl. Acad. Sci. U.S.A. 83:857-561 (1986)

15. Yamamoto, S., Ueda, N., Yakoyama, C., Fitzsimmons, B.J., Rokach, J., Oates, J.A., and Brash, A.R. FASEB Symposium on "Lipoxin: Biosynthesis and Pharmacology, Washington, D.C., March 30 (1987)

16. Lam, B., Marcindiewicz, E., and Wong, P.Y-K. Transformation of 15-hydroperoxyeicosapetaenoic acid into mono- and dihydroperoxyeicosapentaenoic acids by human platelets. In: Drugs Affecting Leukotrienes and Other Eicosanoid Pathways. Edited by B. Samuelsson, F. Berti, G.C. Folco, and G.P. Velo. Plenum Publishing Corp., pp. 167-180 (1987)

6

THE TOTAL SYNTHESIS OF THE LIPOXINS AND RELATED COMPOUNDS

S.E. Webber, C.A. Veale and K.C. Nicolaou

Department of Chemistry
University of Pennsylvania
Philadelphia, PA 19104 USA

INTRODUCTION

In 1984, a new class of arachidonic acid derived products was discovered by Serhan et al.[1-3] While studying lipoxygenase pathway interactions, these researchers isolated two major constituents from human leukocytes. These substances displayed interesting biological properties including the ability to suppress the cytotoxic activity of natural killer cells.[1b,1c,3-7]

The basic structures of these unique eicosanoids were elucidated by use of the standard techniques previously applied in the identification of the eicosanoids. UV spectroscopy indicated that the compuonds possessed conjugated tetraenes (lambda max (MeOH) = 301 nm), while GC-MS analysis established the molecular weight and positions of three hydroxyl groups in each compound. Individual degration experiments were also employed and revealed the chirality of two of the hydroxy groups. The general structures of these two products were assigned as 5,6,15(S)-trihydroxy-7,9,11,13-eicosatetraenoic acid and 5(S),14,15-trihydroxy-6,8,10,12-eicosatetraenoic acid, and were given the trivial names of lipoxin A_4 (LXA_4) and lipoxin B_4 (LXB_4) respectively (Scheme I).[1-3]

Since only minute amounts of the materials are available from biological sources, total synthesis of these compounds was deemed essential. The complete stereochemistry of the hydroxyl groups and the geometries of the conjugated tetraenes could thus be established by direct comparison with the natural samples. In addition, further biological testing of the lipoxins as well as other isomers and analogs would be made possible.

Scheme I. Gross structures of lipoxins A_4 (LXA_4) and B_4(LXB_4).

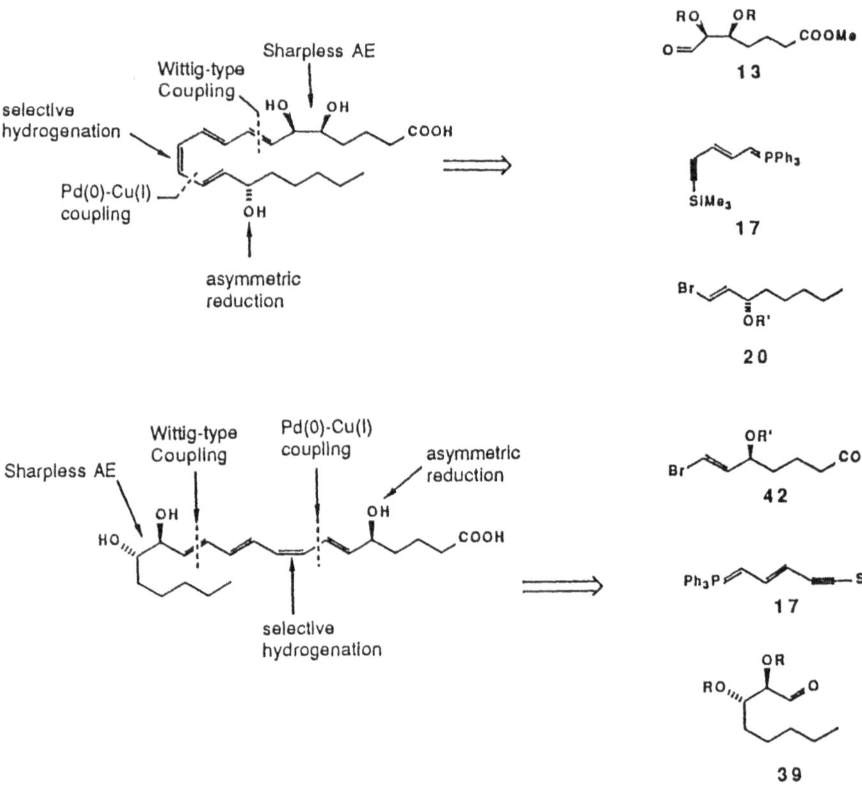

Scheme II. Retrosynthetic analysis of LXA₄ and LXB₄

Due to the potential biological significance[4-7] of the lipoxins, several synthetic projects were initiated by various research groups including our own.[8-11]

GENERAL SYNTHETIC STRATEGY

Since several isomers (stereochemical and geometrical) of LXA₄ and LXB₄ can be generated from biological sources and because the chirality of the hydroxyl groups at C-6 of LXA₄ and C-14 of LXB₄ remained to be determined, we sought to design a general and flexible strategy that would allow us to produce, stereoselectivity, any lipoxin isomer for comparisons with natural samples and for further biological investigations. A retrosynthetic analysis of both LXA₄ and LXB₄ is shown in scheme II. In this analysis each molecule is disconnected into three fragments. Connection of these fragments would utilize both Wittig and Pd(0) coupling[12,13] reactions and would allow predictable construction of the geometry of the tetraene skeleton. The three hydroxyl stereocenters would be set stereospecifically via asymmetric epoxidations[14] and ketone reductions.[15] A final step in each synthetic sequence would be a selective Lindlar hydrogenation of an acetylene generating the cis-tran portion of the tetraene skeleton.

THE TOTAL SYNTHESIS OF LIPOXIN A₄

The synthesis of the LXA₄ and its isomers begins with a four step preparation of the 2,3 epoxy alcohols 4a-d as shown in Scheme III. The stereochemistry of the epoxides were controlled using cis or trans allylic alcohols and Sharpless asymmetric epoxidation[14] techniques. By

utilizing this procedure, the different C-5 and C-6 isomers of lipoxin
A_4 could be prepared selectively depending upon the choice of the
epoxy-alcohol precursor. For the synthesis of 5S,6R-lipoxin A_4,
epoxy-alcohol 4c was used and the sequence outlined in Scheme IV was
followed. This alcohol was transformed into phenyl urethane 5, followed
by intramolecular S_N^2 opening of the epoxide using lewis acid
catalysis yielding the 1,2 cyclic carbonate 6 after hydrolytic work-up.
Transesterfication of the carbonate with sodium methoxide in methanol
gave the 1,2,3-triol 7. The primary hydroxyl of 7 was selectively
protected as the pivolate ester 8, followed by silylation to intermediate
9. Catalytic hydrogenolysis of 9 afforded alcohol 10, which was oxidized
to the crystalline carboxlic acid 11. Reductive cleavage of the
pivaloate ester 11, and diazomethane treatment furnished compound 12,
which was oxidized to aldehyde 13 using PCC.

The second key intermediate, phosphonium bromide 17 was prepared in
multigram quantites from commercially available (2E)-penten-4-yn-1-ol, as
displayed in Scheme V.

The construction of the C-13 to C-20 portion of LXA_4, which
contains the 15(S)-hydroxyl is shown in Scheme VI. This fragment has
been previously used in the synthesis of 5,15 and 8,15 DHETE.[16] Acety-
lenic ketone 18, prepared by a Friedel-Crafts acylation of bis-
(trimethylsilyl)acetylene with hexanoyl chloride, was selectively reduced
to the corresponding (S)-alcohol in >92%, i.e. using (-)-pinanyl-9-BBN,
and the terminal acetylene generated yielding the known (S)-2-octyn-ol

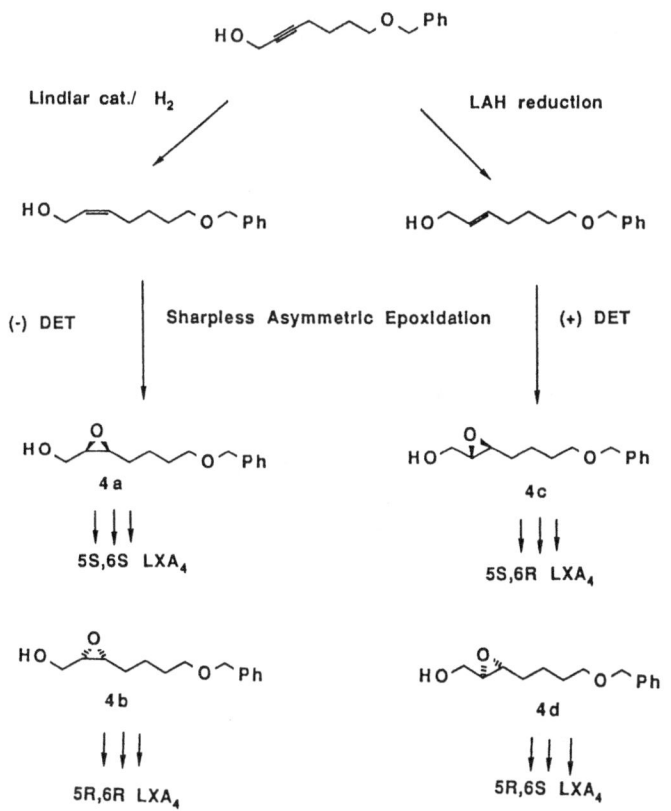

Scheme III. General strategy for the asymmetric construction of lipoxin A_4 (LXA_4).

63

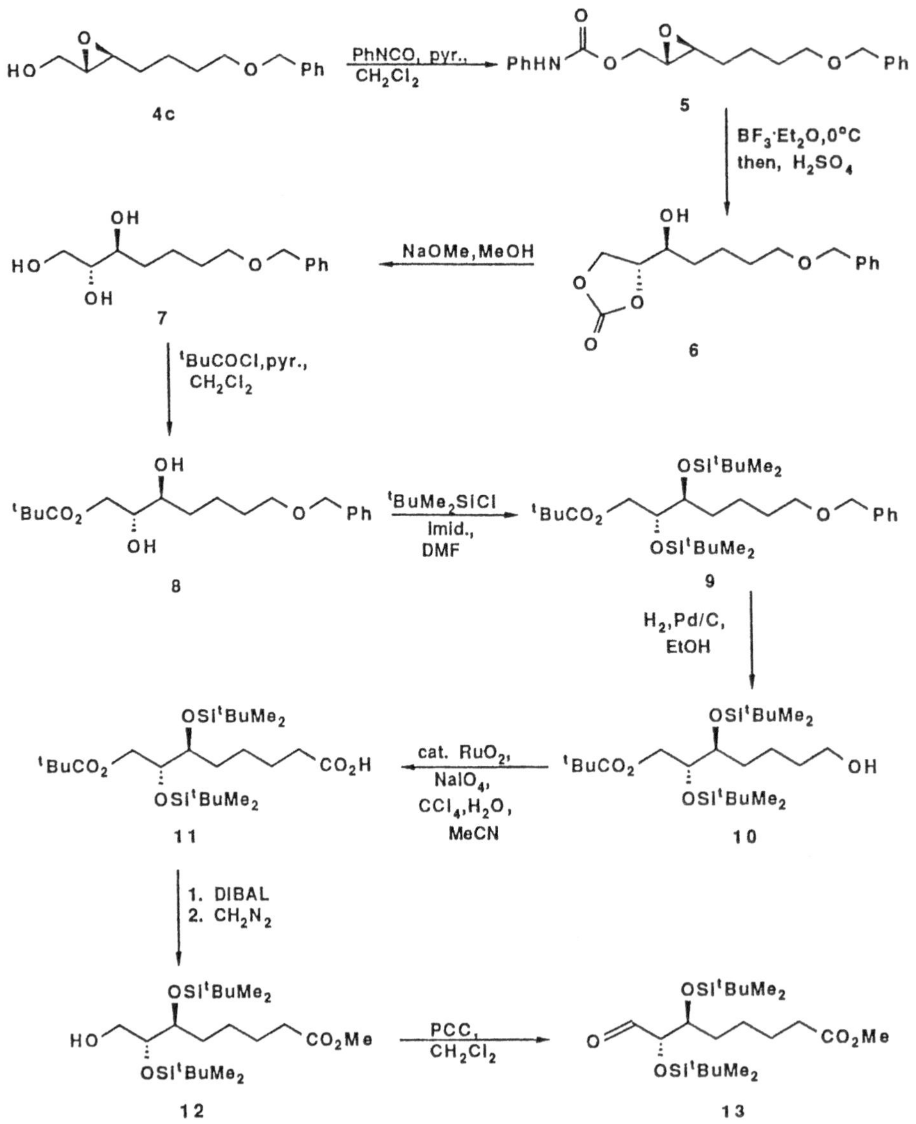

Scheme IV . Synthesis of key intermediate aldehyde 13.

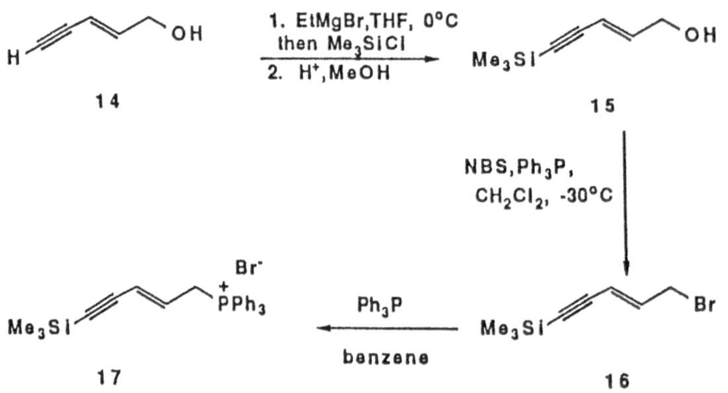

Scheme V. Synthesis of key intermediate phosphonium salt 17 .

64

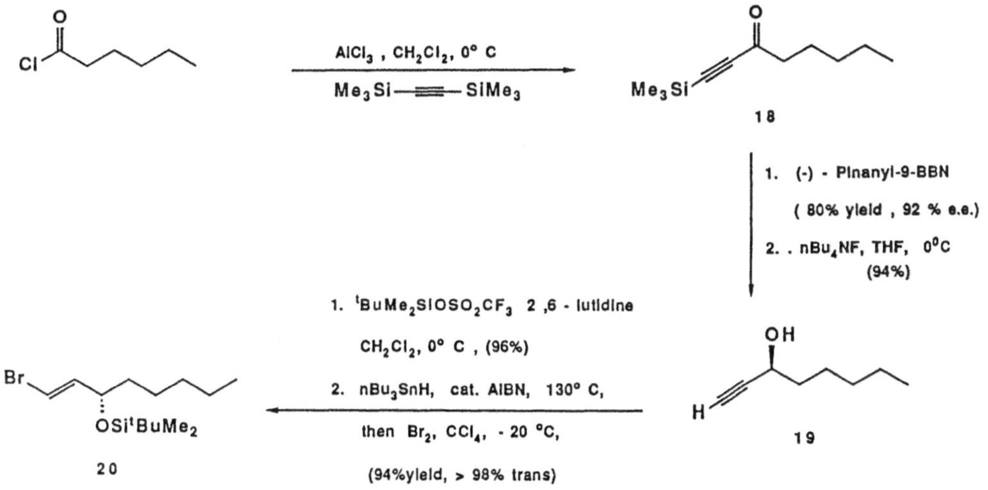

Scheme VI. Synthesis of key intermediate 20.

19. Protection of the alcohol as its silyl ether, followed by hydro-stannylation and bromination afforded trans vinyl bromide 20 (≥98% trans).

The coupling of the major building blocks generating the skeleton of LXA is shown in Scheme VII. The phosphorane of salt 17 was condensed with aldehyde 13, leading to excellent yields of E,Z and E,E-dien-ynes 21 and 22, respectively. The geometrical outcome of the Wittig coupling was poor (Z:E ca 95:5), but it was found that catalytic amounts of iodine could isomerize the E,Z isomer to predominately the E,E-isomer 22 (Z:E ca 10:90). Compound 22 was then purified by chromatography and the E,Z-isomer 21 was recycled. Trans, trans-dienyne 22 was selectively desilylated at the alkyne terminus using AgNO₃, furnishing 23, which was coupled with vinyl bromide 20 by the action of catalytic amounts of Pd(PPh₃)₄ and CuI. The complete skeleton of LXA₄, 24, was formed

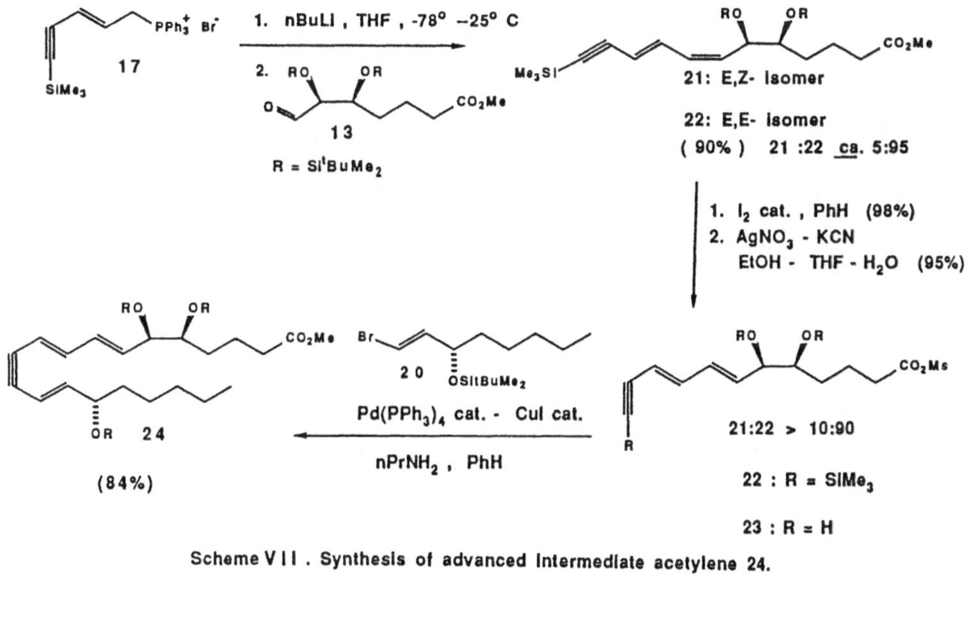

Scheme VII. Synthesis of advanced intermediate acetylene 24.

65

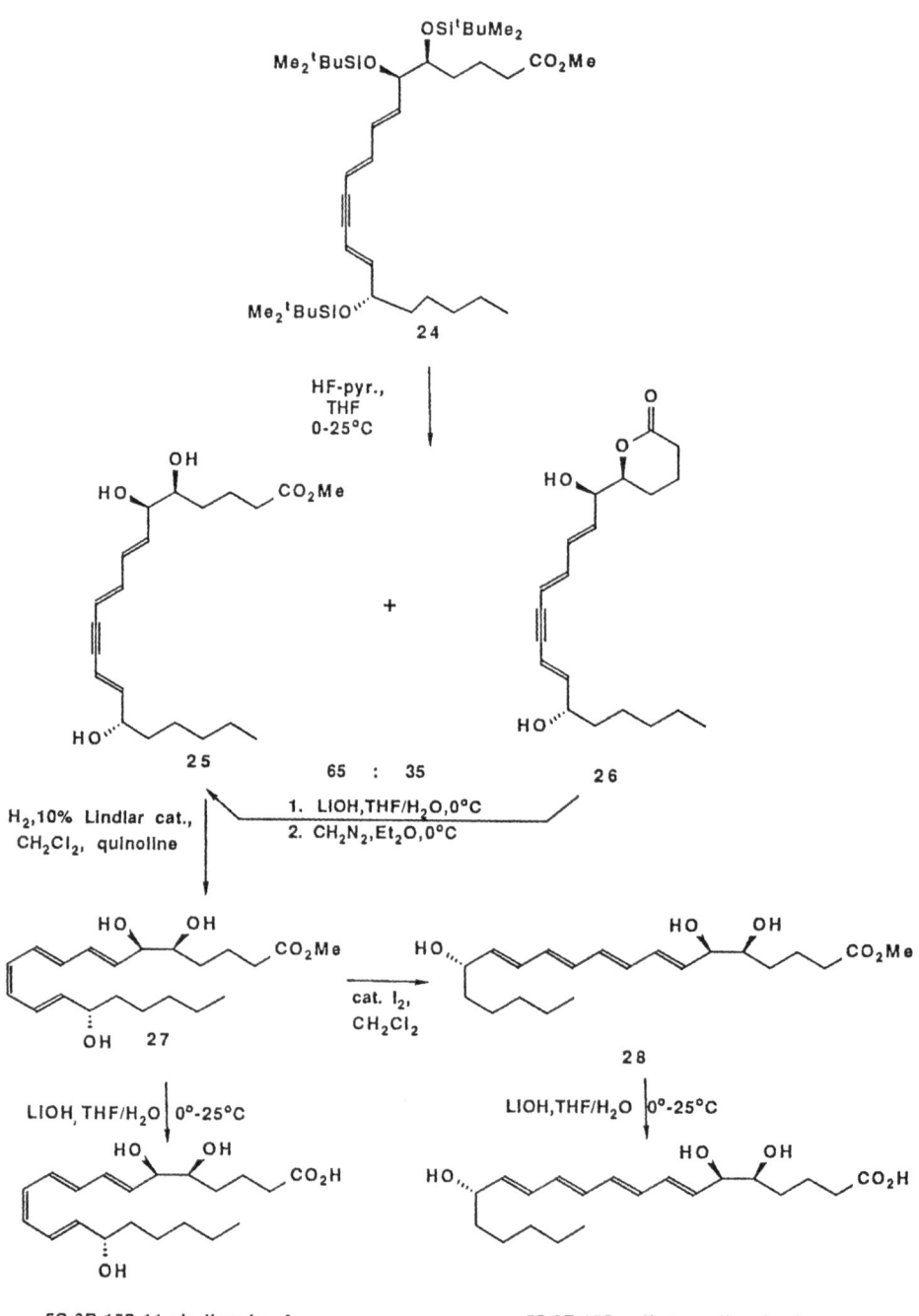

Scheme VIII. Synthesis of lipoxin A₄ and 11-trans-lipoxin A₄.

in this step with retention of the geometrical integrity of the olefin. Desilylation of 24 with hydrogen fluoride-pyridine complex supplied a 65:35 ratio of 25 and lactone 26 (Scheme VIII). The lactone 26 was saponified to the 11,12-dehydrolipoxin A$_4$, which could be esterified to 25 with diazomethane. In the final steps, the acetylene of 25 was selectively reduced to a 5:1 mixture of 5(S),6(R),15(S)-11-cis-LXA$_4$-methyl ester 27 and 5(S),6(R),15(S)-all trans-LXA$_4$-methyl ester 28, which were purified by reverse-phase HPLC. Compound 27 was clearly converted to 28 by exposure to catalytic amounts of iodine. Alkaline hydrolysis of 27 and 28 afforded the corresponding 5(S),6(R) isomers of LXA$_4$. Following a similar sequence but using the epoxy-alcohols 4a and 4b the 5(S),6(R),15(S)-11-cis-LXA$_4$ (6S-LXA$_4$), 5(R),6(R),15(S)-11-cis-LXA$_4$ and their all-trans isomers have also been prepared.

THE TOTAL SYNTHESIS OF LIPOXIN B$_4$

A strategy similar to that used for the LXA$_4$ synthesis was used in the preparation of lipoxin B$_4$ and related isomers. In this strategy (Scheme II) lipoxin B$_4$ is disconnection into three key building blocks; aldehyde 39, phosphonium salt 17, and vinyl bromide 42. Assemblage of these intermediates would provide the 5(S),14(R),15(S)-8-cis, and the 5(S),14(R),15(S)-all trans isomers of lipoxin B$_4$. The epoxy-alcohols 32a-d required for the preparation of various lipoxin B$_4$ isomers were prepared as shown in Scheme IX. The 2(S),3(S)-epoxy alcohol 32d was

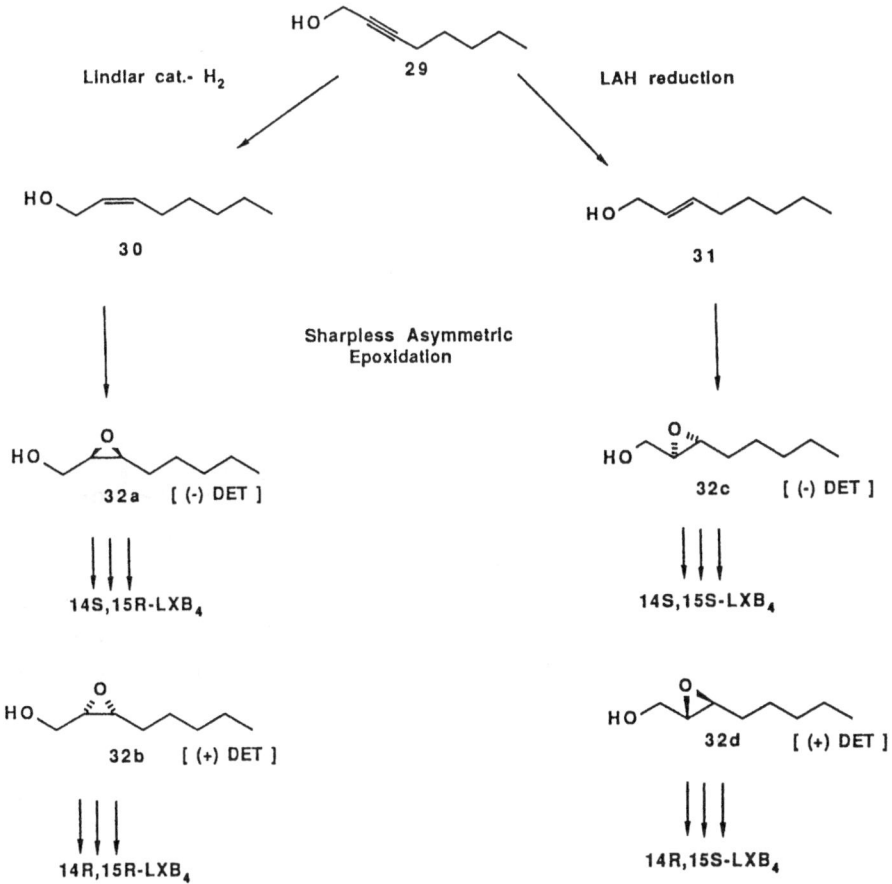

Scheme IX . General stratagy for the asymmetric construction of lipoxin B$_4$ (LXB$_4$).

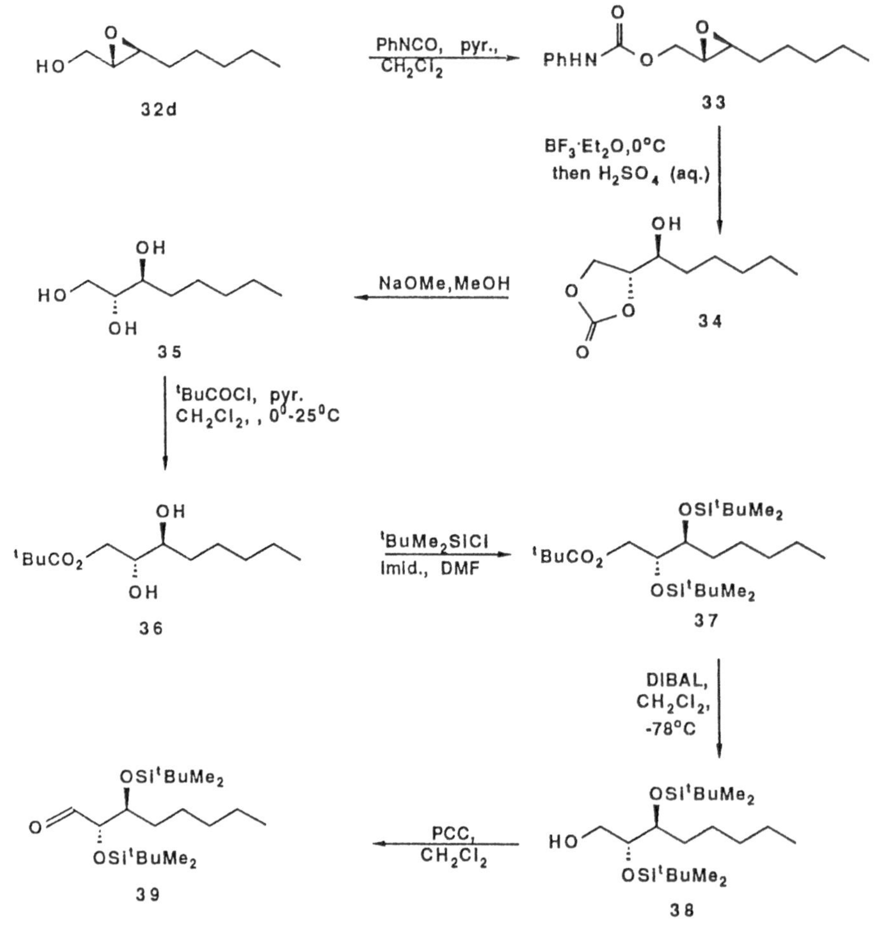

32d

PhNCO, pyr.,
CH₂Cl₂ →

33

BF₃·Et₂O, 0°C
then H₂SO₄ (aq.)

34

NaOMe, MeOH ←

35

^{t}BuCOCl, pyr.
CH₂Cl₂, , 0°-25°C

36

^{t}BuMe₂SiCl
Imid., DMF →

37

DIBAL,
CH₂Cl₂,
-78°C

38

PCC,
CH₂Cl₂ ←

39

Scheme X . Synthesis of key intermediate aldehyde 39.

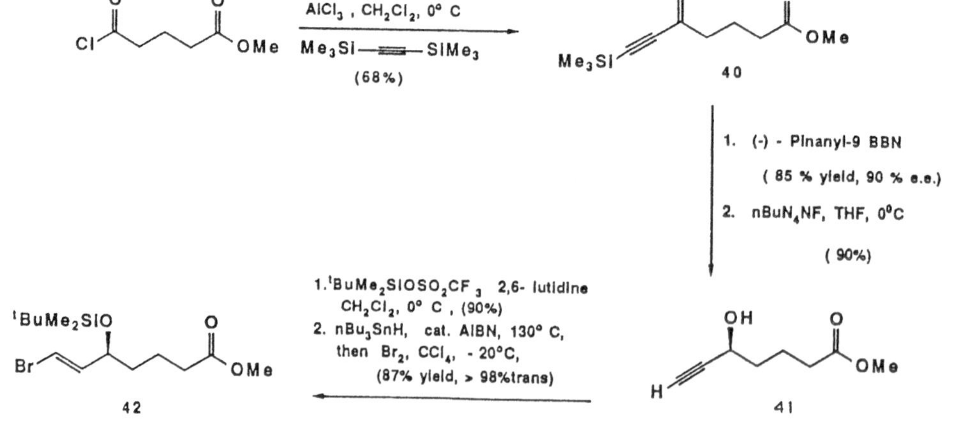

AlCl₃ , CH₂Cl₂, 0° C
Me₃Si———SIMe₃
(68%) →

40

1. (-) - Pinanyl-9 BBN
   (85 % yield, 90 % e.e.)
2. nBuN₄NF, THF, 0°C
   (90%)

41

1. ^{t}BuMe₂SiOSO₂CF₃ 2,6- lutidine
   CH₂Cl₂, 0° C , (90%)
2. nBu₃SnH, cat. AIBN, 130° C,
   then Br₂, CCl₄, - 20°C,
   (87% yield, > 98%trans) ←

42

Scheme XI . Synthesis of key intermediate 42.

68

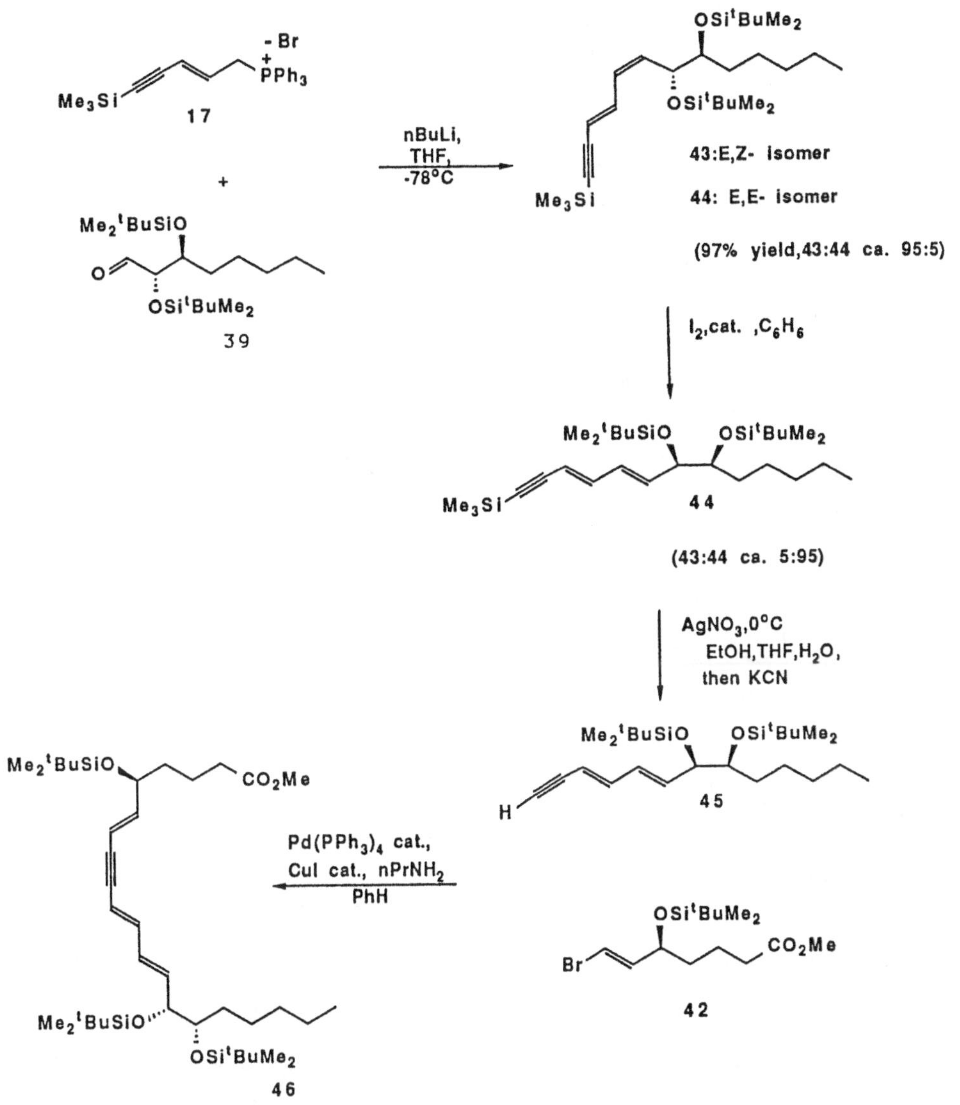

Scheme XII. Synthesis of advanced intermediate acetylene 46.

converted into aldehyde 39 in seven steps as shown in Scheme X. Vinyl bromide 42 (Scheme XI) was generated by an analogous sequence to that used for vinyl bromide 20. We have also used this intermediate in the synthesis of LTB$_4$.[19]. Wittig coupling (Scheme XII) of aldehyde 39 and the phosphorane derived from phosphonium salt 17 gave a mixture of the E,Z and E,E isomers 43 and 44. This reaction as with LXA$_4$ series provided the E,E isomer 44 which, after selective desilylation of the acetylene, was coupled with vinyl bromide 42 to give the acetylene 46. Desilylation gave a mixture of methyl ester 47 and lactone 48. Compound 48 could be converted into 47 via the same sequence as used in LXA$_4$ synthesis. Selective Lindlar hydrogenation gave, after HPLC separation, lipoxin B$_4$ methyl ester which was then converted into 5(S),14(R),15(S)-8-cis-LXB$_4$ (LXB$_4$) and 5(S),14(R),15(S)-all trans-LXB$_4$ (11-trans LXB$_4$) (Scheme XIII).

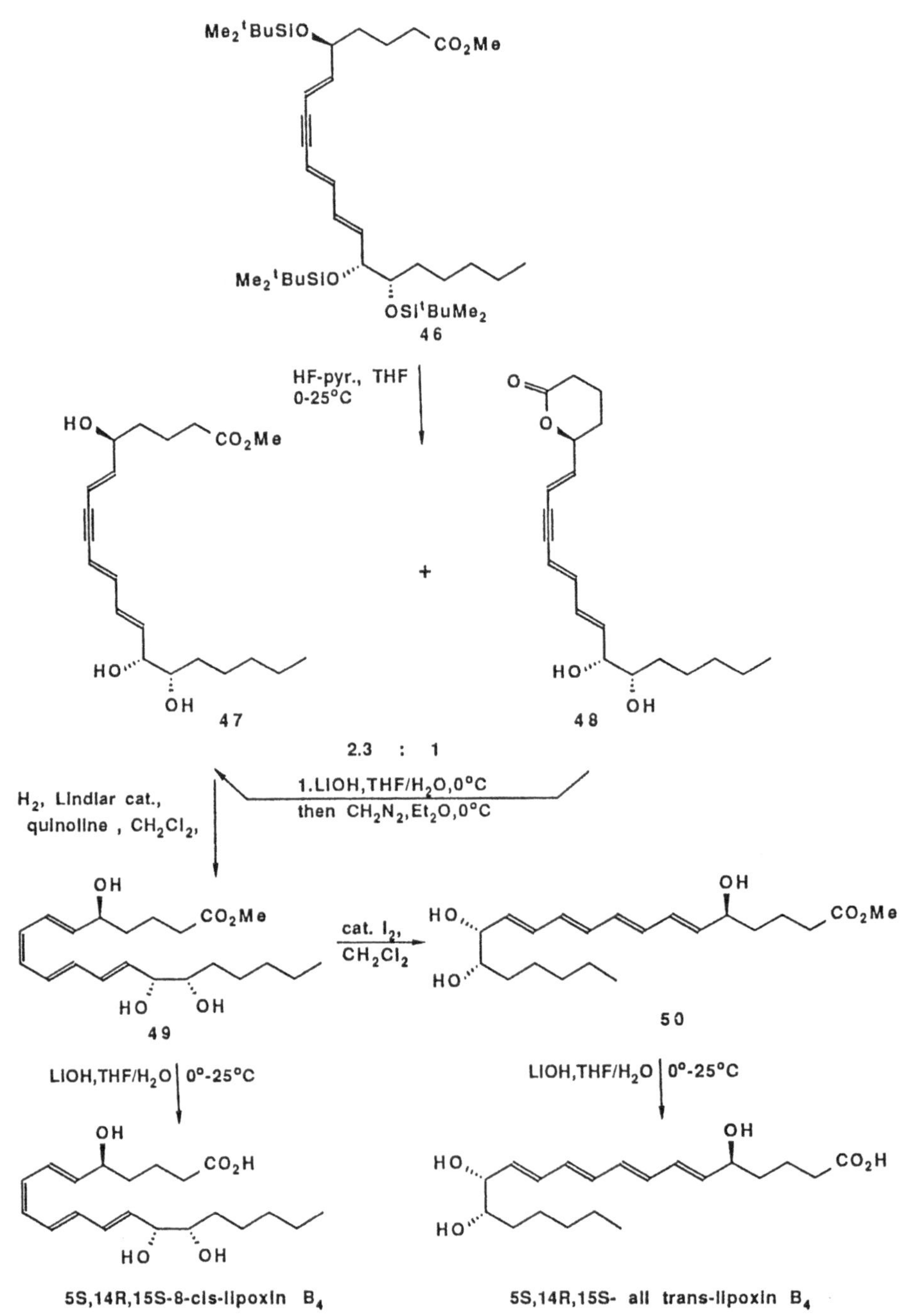

Scheme XIII. Synthesis of lipoxin B_4 and its 8- trans isomer.

THE TOTAL SYNTHESIS OF 5(S),6(S)-EPOXY,15(S)-HYDROXY-7,9,13-<u>TRANS</u>-11-<u>CIS</u>-EICOSATETRAENOIC ACID AND ITS 5,6-METHANO ANALOG

It has been suggested that both lipoxin A_4 and B_4 could arise from a common epoxide intermediate[4] (Scheme XIV). This epoxide could be hydrolyzed either enzymatically or non-enzymatically at the allylic C-6 or C-14 positions generating LXA$_4$ and LXB$_4$, respectively. In

70

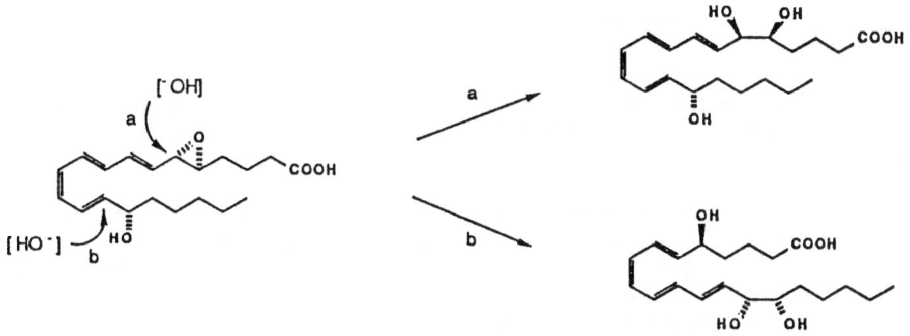

Scheme XIV. Proposed biosynthetic scheme for lipoxin A$_4$ and lipoxin B$_4$

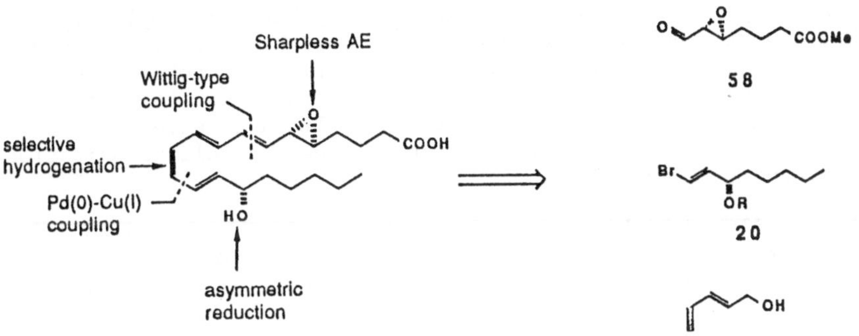

Scheme XV. Retrosynthetic analysis of 5(S),6(S)-epoxy, 15(S)-Hydroxy
-7,9,13-trans-11-cis-eicosatetraenoic acid

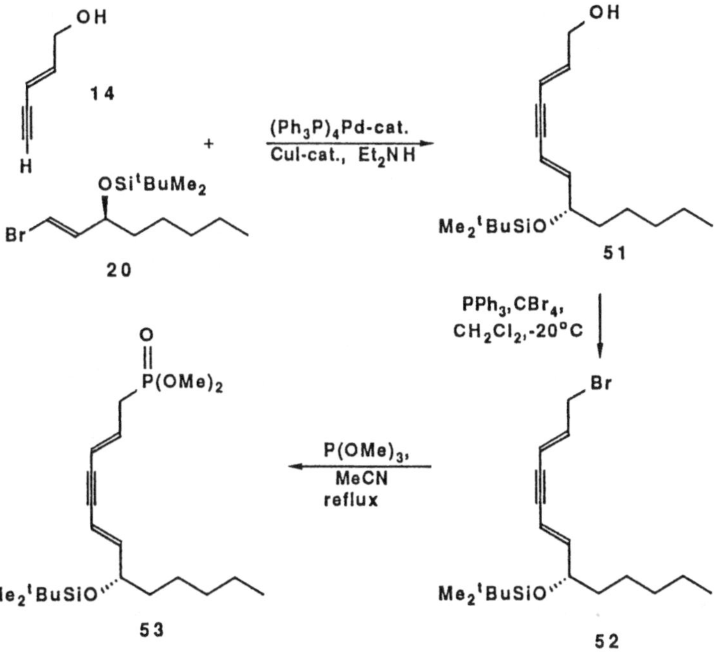

Scheme XVI. Synthesis of key intermediate 53.

order to test this biosynthetic hypothesis we have synthesized this labile epoxide.

As with the other lipoxins this epoxide could be disconnected into three key intermediates as shown in the retrosyntheticd analysis in Scheme XV.

The synthesis of the phosphonate 53 begins with a Pd(o) catalyzed coupling of vinyl bromide 20 (used in the LXA$_4$ synthesis) and commercially available (E)-1-hydroxypent-2-en-4-yne to yield the primary alcohol 51. This alcohol was converted to the allylic bromide 52 and subsequently to phosphonate 53 (Scheme XVI).

Although there are many routes to the widely used epoxy-aldehyde intermediate 58, we were prompted to start wth epoxy-alcohol 4c (Scheme XVII), since it was readily available from the synthetic LXA$_4$ sequence. The primary hydroxyl was protected as the tert-butyldiphenyl silyl ether 54. The benzyl ether was then cleaved by rapid hydrogenolysis using Pearlman's catalyst (Pd(OH)$_2$) producing the primary alcohol 55 with minimal hydrogenolysis of the epoxide. Compound 55 was catalytically oxidized with RuO$_4$, and the resultant carboxylic acid esterified to give 56. Fluoride induced desilylation followed by Collins oxidation generated epoxy-aldehyde 58.

The stabilized lithio-anion of phosphonate 53 was generated and reacted with 58 yielding a 3:1-E,Z mixture of 59 at $\Delta^{7,8}$ (Scheme XVIII). Brief exposure of 59 to catalytic amounts of iodine improved the ratio to 9:1, and this mixture was then separated by HPLC. Difficulties were encountered when several attempts were made to hydrogenate the acetylene group of 59. Selective hydrogenation of the desilyated material 60 could be accomplished, but various HPLC conditions, including the use of deactivaged silica, always led to decomposition. At this stage it was decided to reprotect the allylic alcohol with an acetate group. Thus, selective hydrogenation of acetate 61 furnished the protected epoxy-tetraene 62 after HPLC purification. This compound was

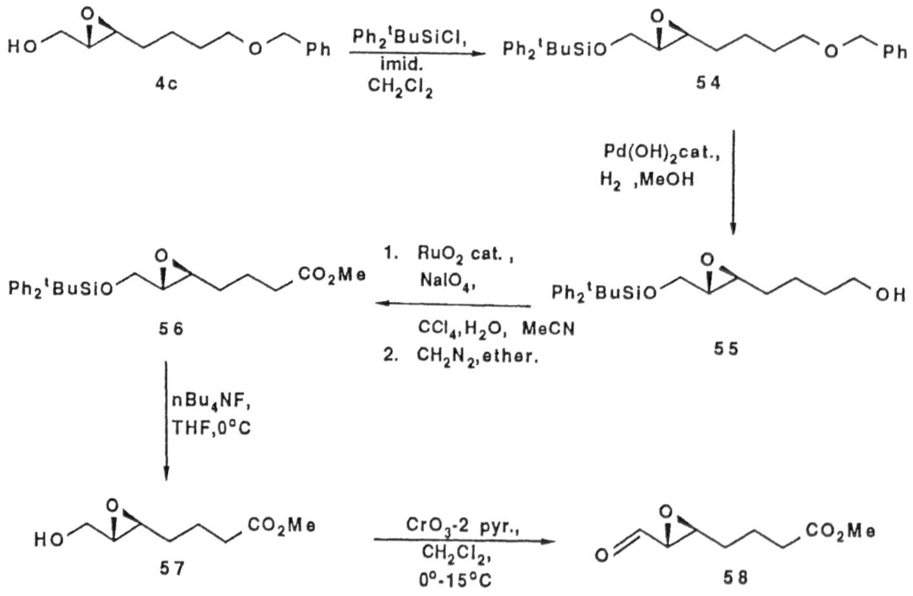

Scheme XVII. Synthesis of key intermediate aldehyde 58.

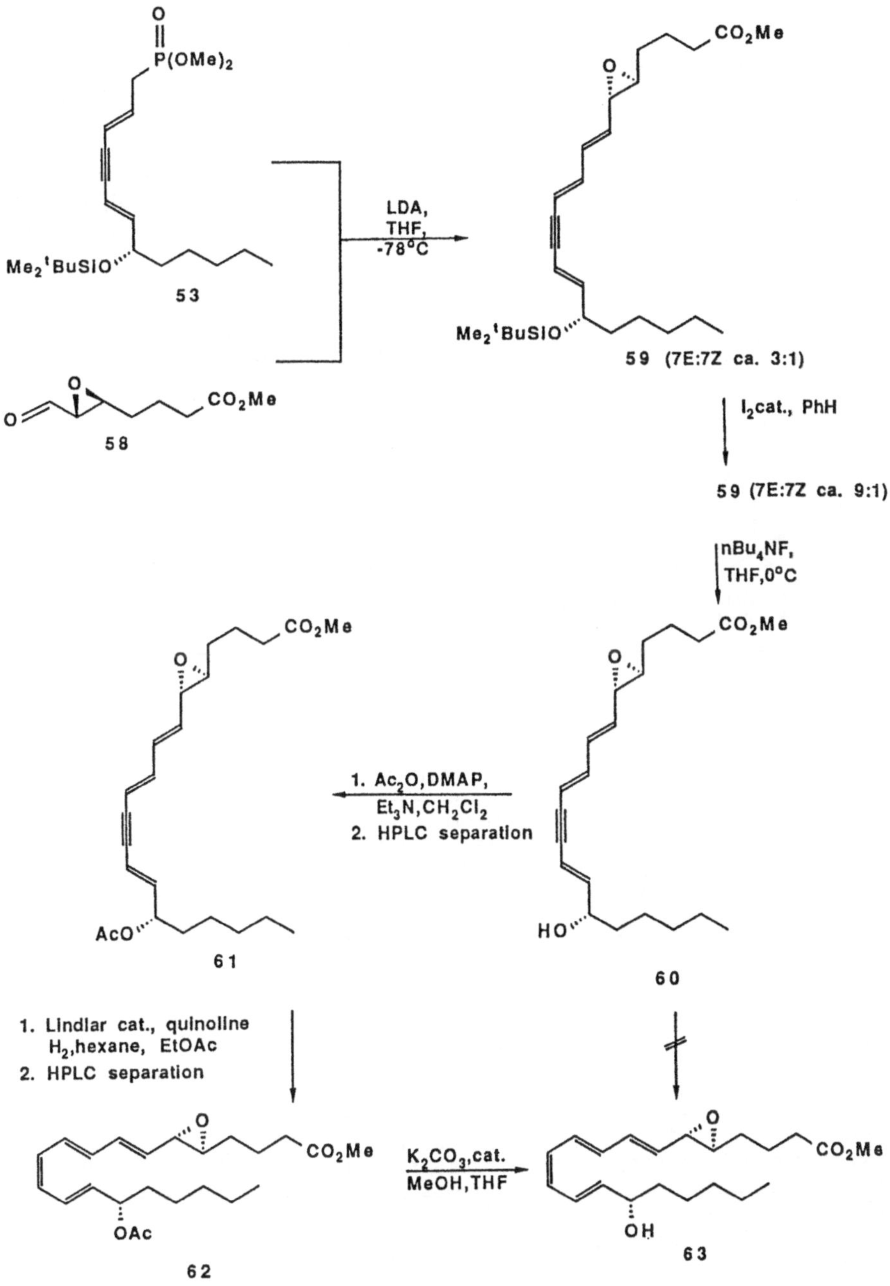

Scheme XVIII . Synthesis of 5,6 epoxytetraene 63.

in turn deacetylated to 63 with catalytic amounts of K_2CO_3 in MeOH. This labile compound can be stored with a trace of ET_3N in benzene at $-40^{\circ}C$ for several days. The methyl ester was saponified under carefully controlled conditions to yield the free acid.

As part of this program we became interested in finding agents to block the biosynthesis of the lipoxins. One idea was to substitute the epoxide moiety of 63 witha cyclopropane. The cyclopropyl group would be roughly isosteric with epoxide, although incapable of undergoing hydrolysis.

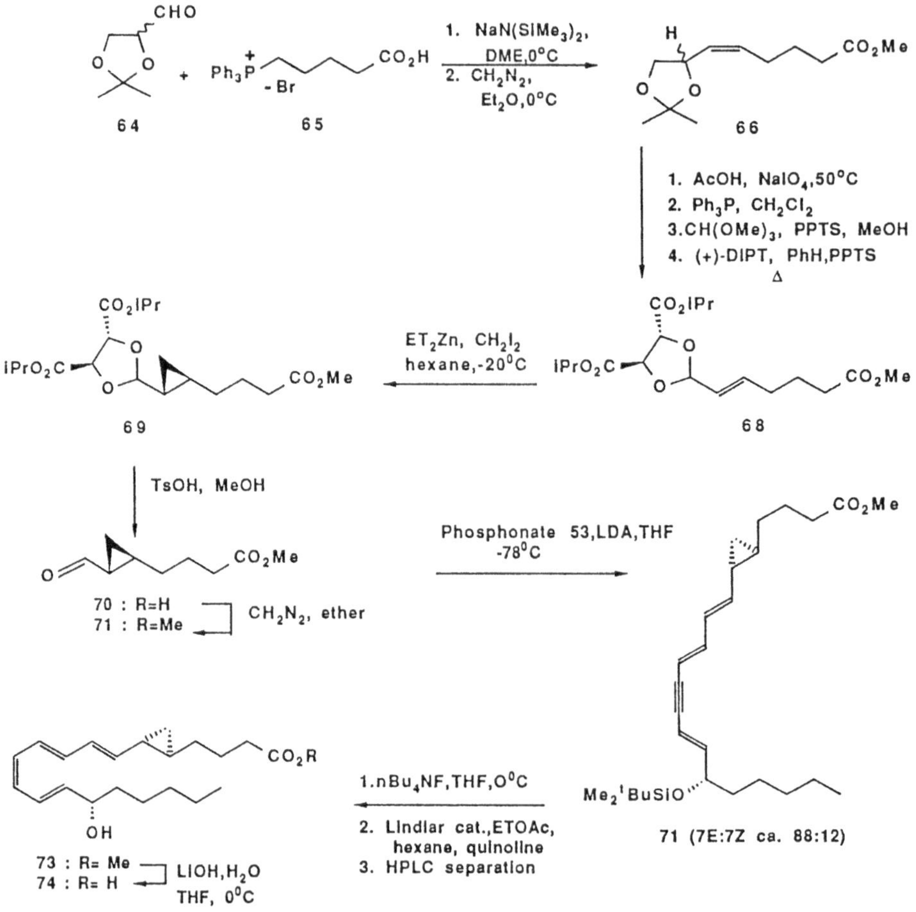

CHO

64 + Ph$_3$P$^+$ ~~~CO$_2$H $\xrightarrow[\text{2. CH}_2\text{N}_2, \text{Et}_2\text{O},0°C}]{\text{1. NaN(SiMe}_3)_2, \text{DME},0°C}$ [structure 66] CO$_2$Me

65

66

1. AcOH, NaIO$_4$,50°C
2. Ph$_3$P, CH$_2$Cl$_2$
3. CH(OMe)$_3$, PPTS, MeOH
4. (+)-DIPT, PhH,PPTS
Δ

CO$_2$iPr [structure 69] CO$_2$Me $\xleftarrow[\text{hexane,-20°C}]{\text{ET}_2\text{Zn, CH}_2\text{I}_2}$ CO$_2$iPr [structure 68] CO$_2$Me

69

68

TsOH, MeOH

[structure 70/71] CO$_2$Me

70 : R=H
71 : R=Me ⎤ CH$_2$N$_2$, ether

Phosphonate 53,LDA,THF
-78°C $\longrightarrow$ [structure 71] CO$_2$Me

Me$_2$^{t}BuSiO''''

71 (7E:7Z ca. 88:12)

[structure 73/74] CO$_2$R

OH

73 : R= Me ⎤ LiOH,H$_2$O
74 : R= H ⎦ THF, 0°C

1.nBu$_4$NF,THF,0°C

2. Lindlar cat.,ETOAc, hexane, quinoline
3. HPLC separation

Scheme XIX . Synthesis of 5,6 methano,15S-hydroxy-7,9,13-trans- 11-cis- eicosatetraenoic acid.

This concept has been used previously, in an analog of leukotriene A$_4$ where it was found that 5,6 methanoleukotriene A$_4$ selectively inhibited the 5-lipoxygenase enzyme.[18,19]

The synthesis of the 5,6-methano analog parallels that of the epoxy compound except that the chiral cyclopropyl-aldehyde 71 would be used in the wittig coupling step. The synthesis begins (Scheme XIX) with a wittig reaction between the known isopropylidene derivative of glyceraldehyde 64 and the ylide of the commercially available (4-carboxybutyl)triphenylphosphonium bromide 65 to give the cis alkene 66. The isopropylidene was hydrolyzed and the resultant diol cleaved by use of NaIO$_4$ in aqueous acetic acid generating the cis-enal, which was then completely isomerized tot he trans-enal by use of triphenylphosphine in methylene chloride. Recently, Yamamoto[20] has developed a procedure was used in generating the chiral cyclopropyl-aldehyde 71. Thus, the dimethyl acetal of the trans-enal was prepared and a trans-acetalation carried out using (+)-diisopropyltartrate. This chiral acetal was then used to direct the subsequent cyclopropanation to give the 5(R),6(R)-methanoacetal. Deprotection of the aldehyde with para-toluenesulfonic acid in methanol also resulted in hydrolysis of the ester giving aldehyde-acid 70, which was then reesterified with diazomethane to yield the desired aldehyde 71. Wittig coupling of this aldehyde with the lithio derivative of phosphonate 53 to give an 88:12, 7E:7Z mixture of acetylenes 71 which were separated by chromatography. Desilylation and

74

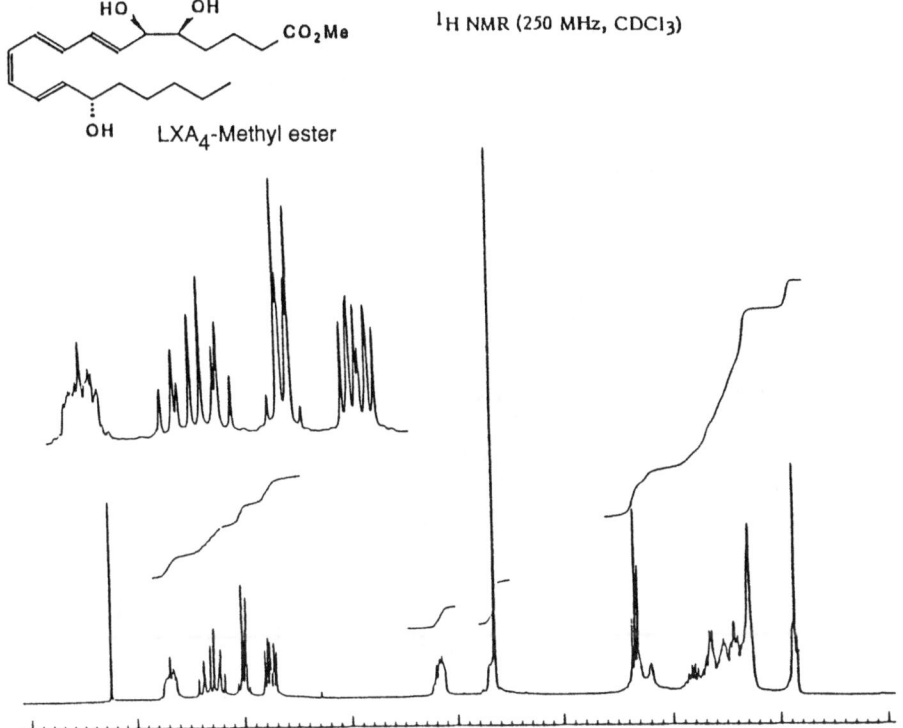

HO OH

CO₂Me

^{1}H NMR (250 MHz, CDCl₃)

LXA₄-Methyl ester

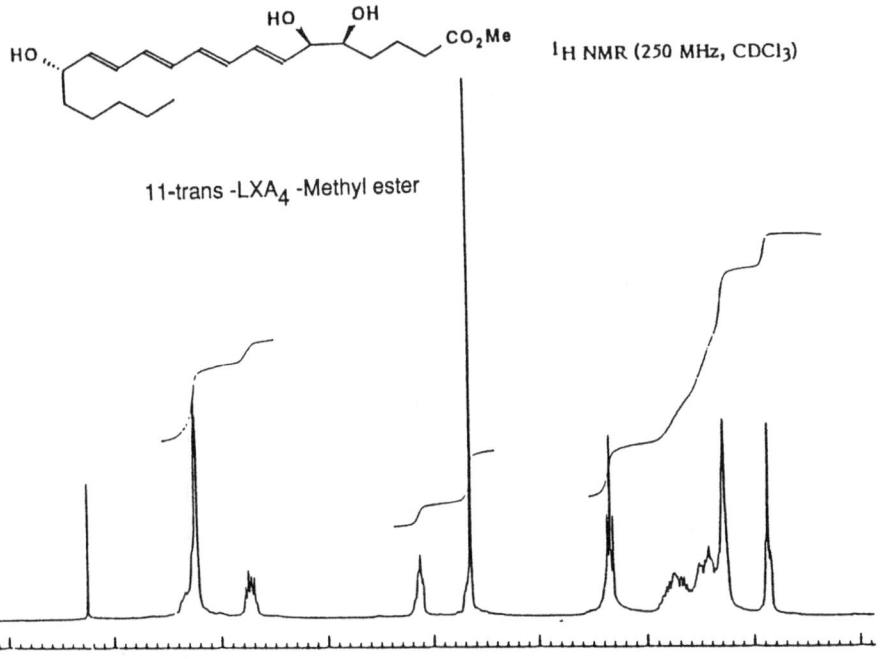

HO OH

CO₂Me

^{1}H NMR (250 MHz, CDCl₃)

11-trans -LXA₄ -Methyl ester

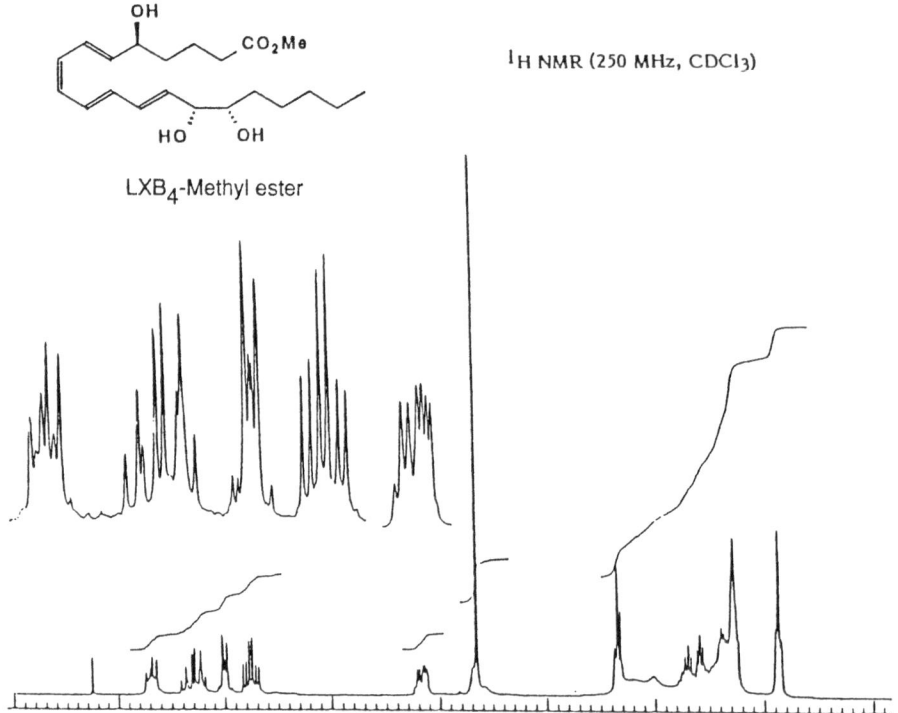

¹H NMR (250 MHz, CDCl₃)

LXB₄-Methyl ester

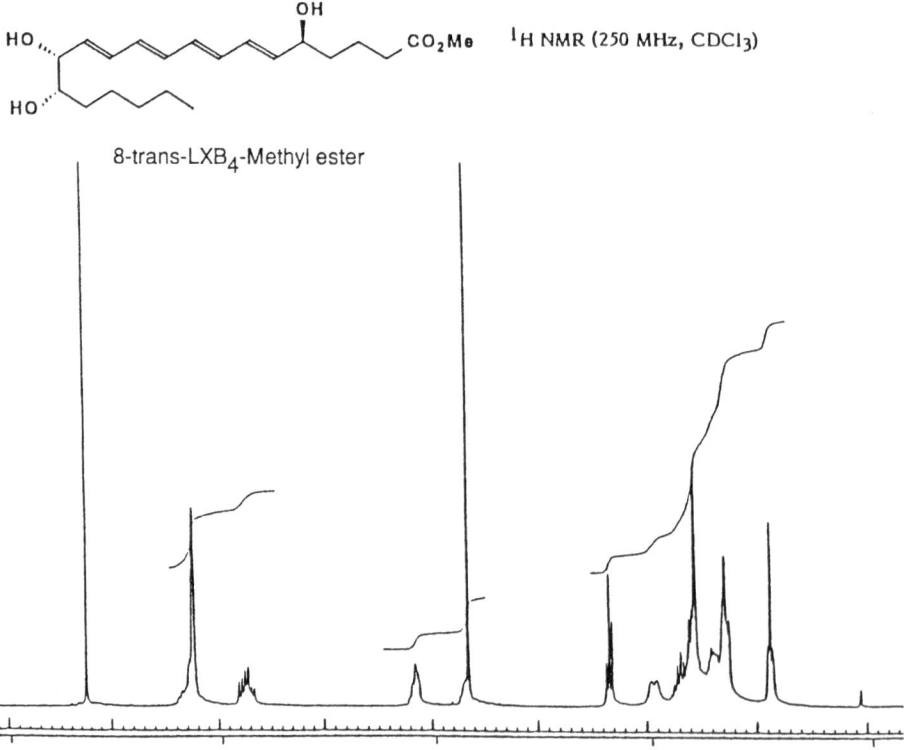

¹H NMR (250 MHz, CDCl₃)

8-trans-LXB₄-Methyl ester

selective hydrogenation were uneventful leading the methyl ester 73, which was saponified to give the 5(R),6(R)-methano-15(S)-hydroxy-11-<u>cis</u>-eicosatraenoic acid 74.

CONCLUSION

In summary, we have synthesized lipoxin A_4 and B_4 as well as several of their isomers by a general strategy based upon Pd(o)-Cu(1) couplings of vinyl halides and terminal acetylenes and selective hydorgenations of the acetylenes to cis double bonds. In addition, we have prepared the proposed intermediate in the biosynthetic sequence, namely the 5,6 epoxy tetraene. These synthetic compounds facilitated in the structural assignments of the natural compounds, and in exploring their biological profiles. Furthermore, the developed strategies are expected to be useful in synthesizing other members of this class of linear eicosanoids.

REFERENCES

1a. B. Samuelsson, S. Hammarstrom, M. Hamberg, and C.N. Serhan, "Advances in Prostaglandin, Thromboxane, and Leukotriene Research," J.E. Pike and D.R. Morton, eds., Raven Press, New York (1985).

1b. "Prostaglandins, Leukotrienes and Lipoxins," J.M. Bailey, ed., Plenum Press, New York (1985).

1c. C.N. Serhan, M. Hamberg, and B. Sameulsson, "Advances in Inflammation Research," F. Russo-Marie, ed., Raven Press, New York (1985).

1d. C.N. Serhan, P. Fahlstadius, S.E. Dahlen, M. Hamberg, and B. Samuelsson, "Advances in Prostaglandin, Thromboxane, and Leukotriene Research," Raven Press, New York (1985).

2. C.N. Serhan, M. Hamberg, and B. Samuelsson, Biochem. Biophys. Res. Commun. 118:943 (1984).

3. C.N. Serhan, M. Hamberg, and B. Samuelsson, Proc. Natl. Acad. Sci. SA 81:5335 (1984).

4. C.N. Serhan, K.C. Nicolaou, S.E. Webber, C.A. Veale, S.E. Dahlen, T.J. Puustene, and B. Samuelsson, J. Biol. Chem. 261:16340 (1986).

5. U. Ramstedt, J. Ng, H. Wigzell, C.N. Serhan, and B. Samuelsson, J. Immunol. 135:3434 (1985).

6. U. Ramstedt, C.N. Serhan, K.C. Nicolaou, S.E. Webber, H. Wigzell, and B. Samuelsson, J. Immunol. 138:266 (1987).

7. A. Hansson, C.N. Serhan, J. Haeggstrom, M. Ingelman-Sundberg, B. Samuelsson, and J. Morris, Biochem. Biophys. Res. Commun. 134:1215 (1986).

8. K.C. Nicolaou, C.A. Veale, S.E. Webber, and H. Katerinopoulos, J. Am. Chem. Soc. 107:7515 (1985).

9. K.C. Nicolaou, and S.E. Webber, J.C.S. Chem. Commun. 5:297 (1985).

10. K.C. Nicolaou, and S.E. Webber, Synthesis :453 (1986).

11. K.C. Nicolaou, and S.E. Webber, J.C.S. Chem. Commun. 24:1816 (1986).

12. K. Songashira, Y. Toda, and N. Hogihara, Tetrahedron Lett. :4467 (1975).

13. V. Ratovelomana, and G. Linstrumelle, Synth. Commun. 11:917 (1981).

14. T. Katsuki, and K.B. Sharpless, J. Am. Chem. Soc. 102:5974 (1980).

15. M. Midland, D.C. McDowell, R.C. Hatch, and A. Tramontano, J. Am. Chem. Soc. 102:867 (1980).

16. K.C. Nicolaou, and S.E. Webber, J. Am. Chem. Soc. 106:5734 (1984).

17. K.C. Nicolaou, R.E. Zipkin, R.E. Dolle, and B.D. Harris, J. Am. Chem. Soc. 106:3748 (1984).

18. K.C. Nicolaou, N.A. Petasis, and S.P. Seitz, J.C.S. Chem. Commun. :1195 (1981).

19. Y. Koshijhara, S. Murota, N.A. Petasis, and K.C. Nicolaou, FEBS Lett. 143:13 (1982).

20. I. Arai, A. Mori, and H. Yamamoto, J. Am. Chem. Soc. 107:8254 (1985).

7

THE LIPOXINS: SYNTHESIS AND BIOSYNTHESIS

Brian Fitzsimmons and Joshua Rokach

Merck Frosst Canada Inc.
P.O. Box 1005
Pointe Claire, Dorval, Quebec
Canada   H9R 4P8

Since the first report of their isolation in 1984[1,2] the lipoxins have generated considerable interest amongst researchers studying the lipoxygenase pathway. The interest has been due both to the novel structure and intreging pharmacological properties[3,4,5] reported for these bio-molecules.

From the synthetic chemist's point of view the lipoxins represented interesting targets for several reasons. First, as is often the case with lipoxygenase products, the quantities available from natural sources is diminishingly small. This limits the amount of structural information that can be obtained using this material to GC/MS and some degradative/GC data which in this case gave the basic connectivity and the stereochemistry of one center. Therefore the double bond geometries and the relative stereochemistry of the asymmetric centers required the availability of structurally unambiguous standards. Second, the lipoxins represent the interaction of different lipoxygenases, if not cell types, thus making the question of their biosynthesis a point of considerable interest. Therefore the synthesis of potential biosynthetic intermediates could provide valuable information about the biosynthesis of the lipoxins. Last, since the lipoxins were reported to possess numerous biological properties, the availability of larger quantities of pure structurally unambiguous material would allow further and more comprehensive study of the activities of these compounds.

The lipoxins were initially described as two peaks on a reverse phaseHPLC trace when the elutant was monitored for absorption at 300 nm.

By a combination of GC/MS and degradative techniques the two basic
structures lipoxin A (LX-A) and lipoxin B (LX-B) were determined (Figure
1). However, neither the geometries of the double bonds nor the relative
stereochemistry of the asymmetric centers were determined, due to the
small quantities isolated. Therefore there was a need for synthetically
prepared standards to complete the structural elucidation of these
compounds.

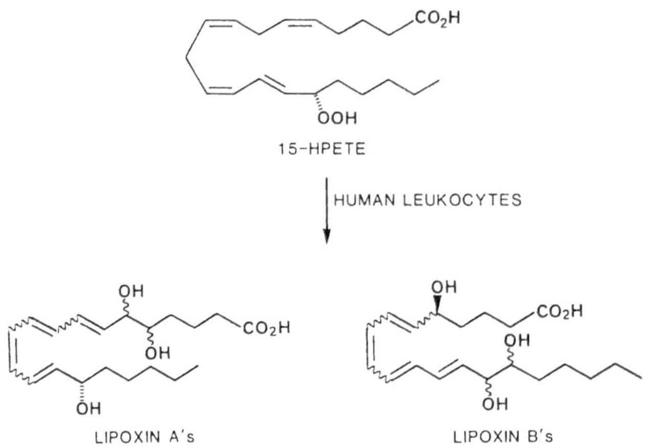

FIGURE 1

Since there were one hundred and twenty-eight possible structures for
the combination of lipoxin A and lipoxin B, the preparation of all the
possible isomers was an unwieldly task. However, we felt that by
carefully analyzing the biosynthetic possibilities this number could be
substantially reduced. The first transformation the 15-HPETE would be
expected to undergo is lipoxygenation at C5 to give 5S,15S-diHPETE, thus
fixing the C5 stereocenter as S in both lipoxin A and lipoxin B. This is
in fact a well documented conversion and its occurance in this case was
supported by the isolation of substantial quantities of 5,15-diHETE from
the incubation mixture. The 5,15-diHPETE could then suffer one of two
fates. The first possibility was that a further lipoxygenation could
occur at C6 or at C14 to yield a lipoxin A or a lipoxin B respectively.
Based on the result of a similar lipoxygenation of 15-HPETE to give
14,15-diHPETE, the most likely products of this pathway, after reduction
of the hydroperoxides, were felt to be the 5S,6R,15S-11-cis lipoxin A
isomer 2 and the 5S,14R,15S-8-cis lipoxin B isomer 8 (Figure 2). It
is important to note that although the newly created asymmetric center is
defined as having the R absolute configuration it is produced by addition
of oxygen to the same face of the substrate as in the production of
5S-HPETE or 15S-HPETE from arachidonic acid. The only difference being

that the priorities assigned to the bonds of the new asymmetric center are changed by the presence of the neighbouring hydroxy-methine moiety in the case of the lipoxins. We initially tested the predicted outcome of this "triple lipoxygenation" pathway by the incubation of arachidonic acid with soybean lipoxygenase. This incubation yielded the lipoxin A isomer **2** and the lipoxin B isomer **8** thus supporting the prediction that the asymmetric center generated by the final lipoxygenation would be of the R absolute configuration. Subsequently Yamamoto et al. and Kuhn et al. (see respective chapters in this volume) have also demonstrated this outcome for the lipoxygenase pathway of lipoxin biosynthesis.

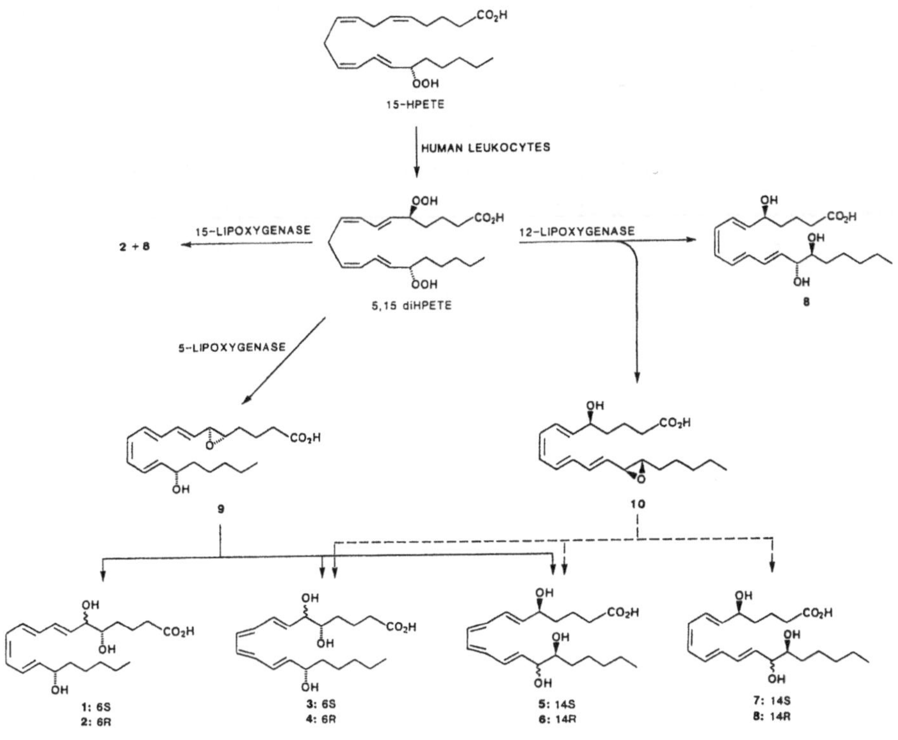

FIGURE 2

The second biosynthetic possibility considered was that the 5,15-diHPETE could undergo one of two stereospecific dehydrations to give the epoxytetraenes **9** and **10**. This would be directly analogous to the formation to the formation of LTA$_4$ from 5-HPETE or 14,15 LTA$_4$ from 15-HPETE respectively. Once formed the epoxides **9** and/or **10** could yield lipoxins by several routes. The most straightforward of these would be vicinal hydrolysis of the epoxide moiety either with inversion of stereochemistry to give the 5S,6R,15S-11-cis lipoxin A isomer **2** from the epoxide **9** and the 5S,14R,15S-8-cis lipoxin B isomer **8** from the

81

epoxide 10. Hydrolysis with net retention of configuraton would yield the 5S,6S,15S-11-cis lipoxin A isomer 1 and the 5S,14S,15S-8-cis lipoxin B isomer 7 from 9 and 10 respectively. These epoxides would also be expected to give rise to the all trans lipoxin A's 3 and 4 and lipoxin B's 5 and 6 by aqueous hydrolysis, analogously to the formation of 6-trans and 6-trans, 12-epi LTB_4 from LTA_4.

While enzymatic conjugate hydrolysis, as in the conversion of LTA_4 to LTB_4, was considered as a possibility, the lack of precedents made any predictions of the outcome of such a process quite uncertain.

Therefore we had paired the original 128 possible isomers down a much more managable 8. Now that we had out targets it was necessary to lay out the criteria for their total synthesis. First it was felt that to ensure the correct identification of the asymmetric centers in the natural products they should all be obtained directly from carbohydrate precursors. In addition the synthetic plan should allow for the verification that epimerizations had not occurred. Second, since the predicted isomers are all cis/trans pairs, this bond would be the last carbon-carbon link contracted and a method that would give both the cis and trans product would be used.

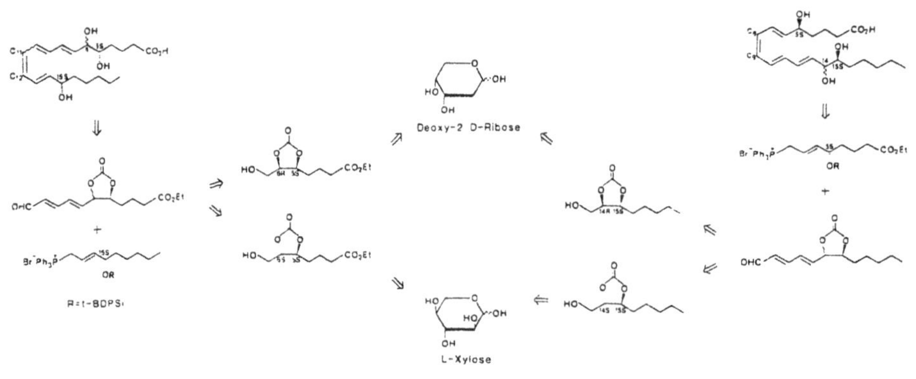

FIGURE 3

Retrosynthetic analysis of the lipoxins (Figure 3) indicated that the 6/14R (lipoxin A/B) diasteriomers could be derived from 2-deoxy-D-ribose while the 6/14S diasteriomers could be prepared from L-xylose.

A cyclic carbonate was chosen as the protecting group for the vicinal diol for several reasons. First it allowed for the protection of two hydroxyl moieties with a single group, secondly it would readily removable under conditions where the final product would be stable. Lastly it imported a rigidity to this portion of th emolecule thussimplifying the verification that epimerization had not occured in any step. While formation of the cyclic carbonate from 2-deoxy-D-ribose could be formed directly, that from L-xylose was envisaged as coming from cyclization of a mixed carbonate at an advanced stage of the synthesis. Also in the case of L-xylose excision of the C-2 hydroxyl group was necessary, this being accomplished by the procedure of Wong and Gray.

The synthesis of the bottom half of the lipoxin A's is shown in Figure 4. Hydrolysis of the thioacetal of the latent dialdehyde 11 followed by Wittig olefination of the aldehyde and then hydrogenation of the olefin gave the acetonide 12. Treatment of the acetonide 12 with aqueous acid gave the corresponding diol which was subjected to lead tetraacetate cleavage to generate the second aldehyde moiety. This aldehyde was homologated with formyl methylidene triphenyl phosphorane to give the α,β unsaturated aldehyde 13. The aldehyde 13 was then converted to the phosphonium salt 14.

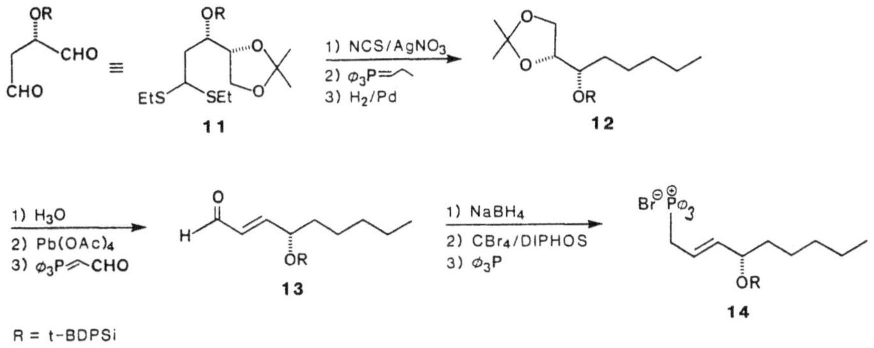

FIGURE 4

The synthesis of the 5(S), 6(R), 15(S) isomers of lipoxin A utilized 2-deoxy-D-ribose as the starting material (Figure 5). Glycosidation of the sugar followed by formation of the 3,4 cyclic carbonate and then hydrolysis of the glycoside yielded the intermediate 15 in which the future 5,6-diol is protected with a single protecting group and the termini are distinguishable. Condensation of the hydroxyaldehyde with

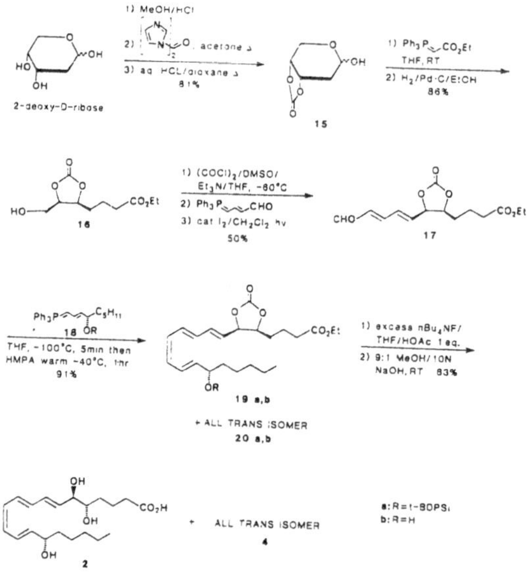

FIGURE 5

ethyl (triphenyl-phosphoranylidene) acetate and hydrogenation of the olefin generated afforded the alcohol **16**. Oxidation of the alcohol **16**, condensation of the aldehyde **17** produced with 4-(triphenyl-phosphoranylidene) but-2-enal, and isomerization provided

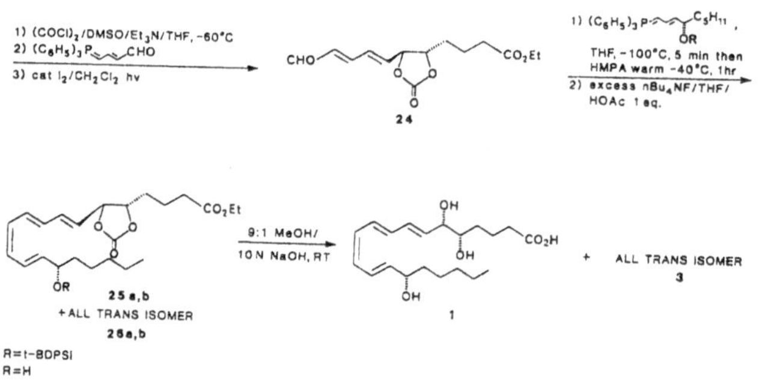

FIGURE 6

the diene aldehyde required for the formation of the final carbon-carbon double bond. Condensation of the aldehyde <u>17</u> with the ylide <u>18</u> generated by the treatment of the phosphonium salt <u>14</u> with 1 equivalent of nBuLi at -100°C for 2 min. yielded the two 5(S), 6(R), 15(S) isomers <u>22a</u> and <u>23a</u> in a 1:1 ratio. Finally removal of the silyl protecting group, separation of the 11-<u>cis</u> and the 11-<u>trans</u> isomers and basic hydrolysis afforded two 5(S), 6(R) isomers of lipoxin A <u>2</u> and <u>4</u>.

L-xylose was chosen as the starting material for the synthesis of the 5(S), 6(S), 15(S) isomers of lipoxin A. Since it was necessary to excise the C-2 hydroxyl group and to protect the C-3 and C-4 hydroxyl groups, a simple and efficient method to prepare the 6(S) diastereomer of the alcohol <u>16</u> was devised. Pivotal to this scheme was the formation of the cyclic carbonate from an acyclic carbonate upon hydrolysis of the acetonide. Using the method Wong and Gray, L-xylose was converted into the alcohol <u>21</u> in four steps (Figure 6). Protection of the alcohol as its ethyl carbonate derivative, hydrolysis of the thioacetal, and condensation of the resulting aldehyde with ethyl (triphenyl-phosphoranylidene) acetate gave the acyclic carbonate <u>22</u>. Hydrogenation gave the corresponding saturated compound. Hydrolysis of the acetonide occurred with concomitant formatin of the cyclic carbonate as expected, yielding the alcohol <u>23</u>. The alcohol <u>23</u> is diastereomeric with the alcohol <u>16</u> and was converted to the 5(S), 6(S), 15(S) isomers of lipoxin A using the same procedure as for the preparation of the 5(S), 6(R) isomers of lipoxin A from the alcohol <u>16</u>.

The C-1 - C-8 fragment common to all of the lipoxin B isomers was prepared as 9 shown in Figure 7. Treatment of the thioacetal <u>27</u>, prepared from D-arabinose, with N-chlorosuccinimide/silver nitrate in acetonitrile-water at -20°C gave the corresponding aldehyde which was condensed with 1 equivalent of (carboethoxymethylene) triphenyl-phosphorane in dichloromethane to yield, after hydrogenation of 5% Pd-C, the ester <u>28</u> in a 71% yield. Hydrolysis of the acetonide of <u>28</u> with trifluoroacetic acid in THF-water, lead tetraacetate cleavage of the resultant diol in dichloromethane at -78°C, followed by condensation of the aldehyde with 1.2 equivalents of (formylmethylene)triphenyl-phosphorane in toluene at 80°C for 5 h gave the α,β unsaturated aldehyde <u>29</u> in a 61% yield. Reduction of the aldehyde <u>29</u> with $NaBH_4$-$CeCl_3$ in isopropanol-water at 0°C, treatment of the resulting alcohol with DIPHOS and carbon tetrabromide in dichloromethane at room temperature gave the corresponding bromide, which upon treatment with

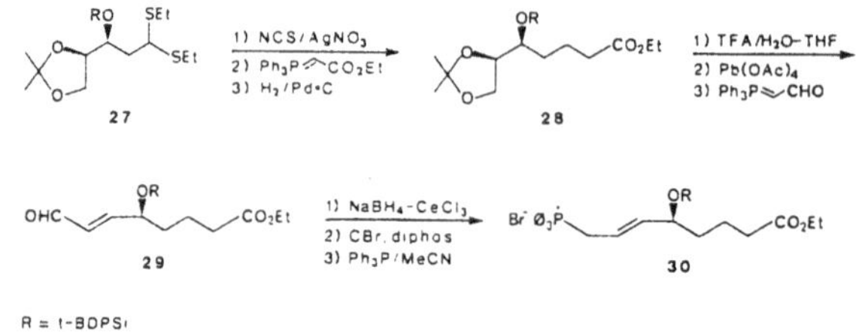

FIGURE 7

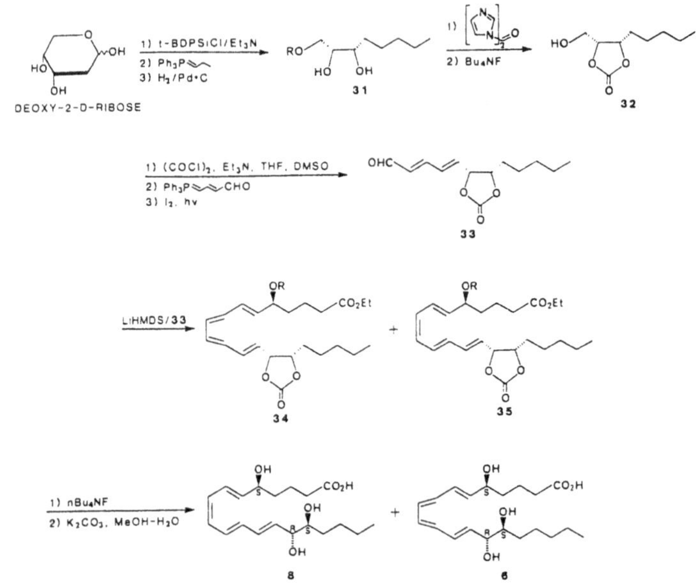

FIGURE 8

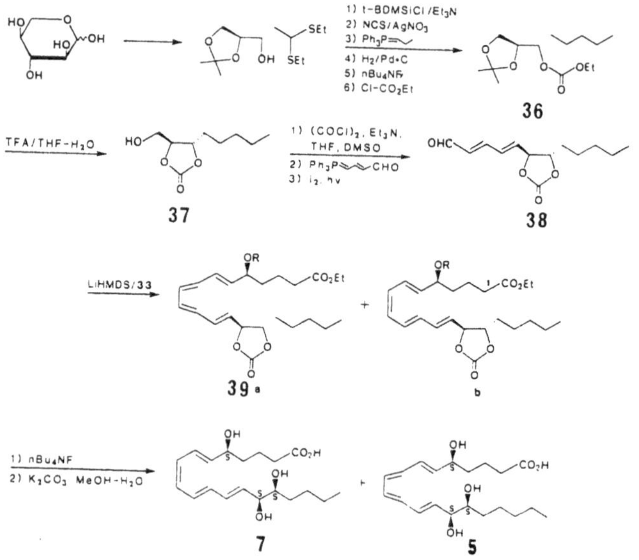

FIGURE 9

86

excess triphenylphosphine in acetonitrile yielded the phosphonium salt 30 in an 86% yield.

The 14R,15S isomers of lipoxin B were prepared as shown in Figure 8. Protection of 2-deoxy-D-ribose with t-butylchlorodiphenylsilane and triethylamine in dichloromethane gave the corresponding 5-O-silyl derivative in an 81% yield. Condensation of the lactol with excess (1-propylene)triphenylphosphorane in THF and hydrogenation of the resulting olefin over 5%-Pd-C gave the diol 31 in a 76% yield. Treatment of the diol 31 with excess 1,1'carbonyldiimidazole in methylethylketone at 100°C in a sealed tube and subsequent removal of the silyl group by treatment with tetra-n-butylammonium fluoride in THF gave the alcohol 32 in a 75% yield. Oxidation of the alcohol, condensation of the resulting aldehyde with $Ph_3P=CH-CH=CH-CHO$ in THF and photolysis of the resulting cis/trans mixture with a catalytic amount of iodine in dichloromethane gave the trans, trans dienealdehyde 33 in a 47% yield. Condensation of the aldehyde 34 at -100°C with 2 equivalent of the ylide generated by treatment of the phosphonium salt 30 with 0.9 equivalents of LiHMDS in THF at -100°C, subsequent addition of HMPA and warming to -40°C for 2 h gave a 1:1 mixture of the two desired protected 14R,15S isomers of lipoxin B 34 and 35 in a 92% yield. The cis and trans isomers were separated easily by HPLC (12.5% EtOAc/hexane). Deprotection of the isolated isomers with tetra-n-butylammonium fluoride in THF and then potassium carbonate in aqueous methanol at 5°C gave the corresponding lipoxin B isomers 6 and 8 in an 83% yield.

The 14S,15S isomers of lipoxin B were prepared as shown in Figure 9. Treatment of the alcohol 35 with t-butylchlorodimethylsilane, DMAP (cat) and triethylamine in dichloromethane at room temperature gave the corresponding silylated alcohol in an 84% yield. Hydrolysis of the thioacetal with N-chlorosuccinimide and silver nitrate in acetonitrile-water at -20°C generated the free aldehyde which was condensed with (1-propylene)triphenylphosphorane in THF at 0°C to afford the corresponding olefin which was hydrogenated over 5% Pd-C to give the saturated compound. The silyl protecting group was then removed by treatment with excess tetra-n-butylammonium fluoride in THF at 0°C. Treatment of the resulting alcohol with ethylchloroformate in pyridine gave the mixed carbonate 36 in a 67% yield from the thioacetal 35. Treatment of the mixed carbonate 36 with trifluoroacetic acid in

THF–water caused hydrolysis of the acetonide with concurrent formation of the cyclic carbonate. The cyclic carbonate 37 was converted into the diene aldehyde 38 by the same sequence of reactions used to convert the alcohol 32 into the diene aldehyde 33 (Figure 8) in a similar yield. Condensation of the aldehyde 38 with the phosphorane generated from the salt 30 as described for the 14R, 15S isomers gave the desired protected 14S, 15S lipoxin B isomers 39a and b in an 87% yield. These isomers were separated and deprotected as described for the 14R, 15S lipoxin B isomers to give the free 14S, 15S lipoxin B isomers 5 and 7 in similar yield.

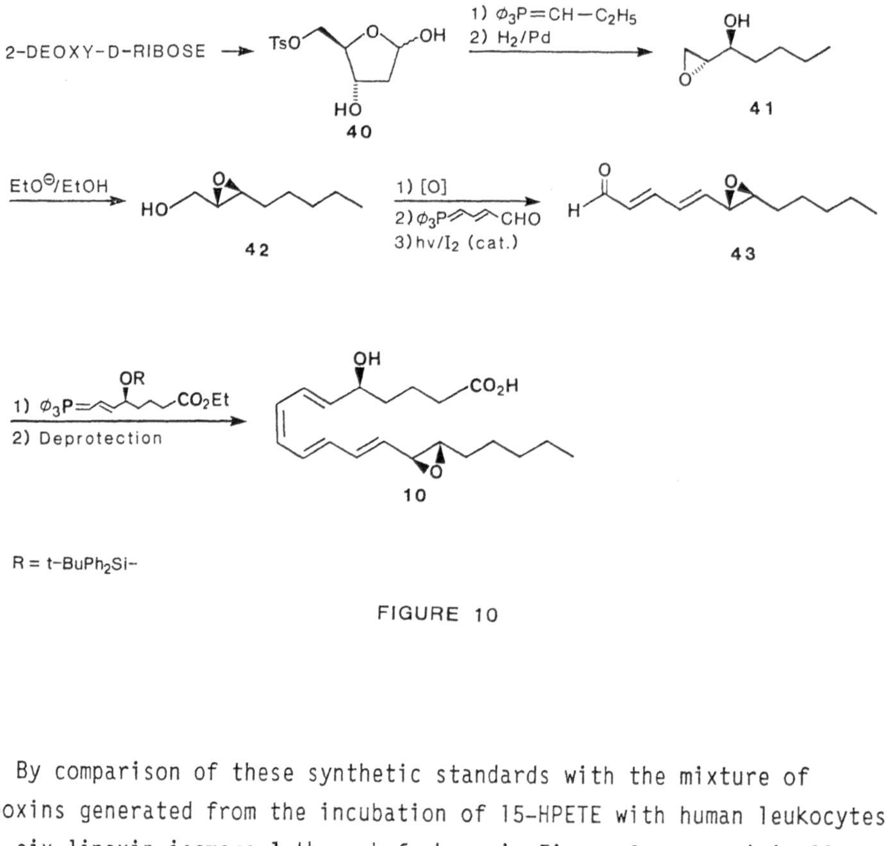

FIGURE 10

By comparison of these synthetic standards with the mixture of lipoxins generated from the incubation of 15-HPETE with human leukocytes the six lipoxin isomers 1 through 6 shown in Figure 2 were originally identified.[7,8,9] These six isomers were shown to be also produced by aqueous hydrolysis of synthetically prepared 5,6-epoxy-15 hydroxy tetraene epoxide 9 postulated as an intermediate in the biosynthesis of the lipoxins.[8] This, along with the isolation of methanol trapping products, lead to the conclusion that the tetraene epoxide 9 was formed biolgocially and these these six lipoxin isomers were derived from it.

Subsequent to this work a seventh isomer has been found by these laboratories and others, that being the 5S, 14R, 15S-8-cis-lipoxin B isomer 8 (Figure 2).[11] The origin of this isomer is at present open to speculation, however it is unlikely that it arises from the 5,6 tetraene epoxide 9 by non-enzymatic hydrolysis, which can account for the six isomers previously described.

Two possible pathways by which this isomer culd be produced are outlined in Figure 2. While the lipoxygenase pathway had been shown to yield the isomer 8 by the action of 12- or 15- lipoxygenase[9,12] on 5,15 diHPETE, the 14,15 tetraene epoxide 10 had not been investigated as a possible precursor.

To prepare the bottom epoxide bearing portion of the target, methyl 2-deoxyribofuranoside was chosen as the starting material. Derivatization of the primary hydroxyl group as its sulfonate ester followed by hydrolysis of the glycoside gave the lactol 40. Treatment of the lactol 40 with propylidenetriphenylphosphorane caused concomitant formation of the olefin and terminal epoxide to give the epoxide 41. Hydrogenation followed by treatment with sodium ethoxide gave the internal epoxide. Oxidation of the primary hydroxyl group and condensation of the resultant aldehyde with 3-formylpropenylidenetriphenylphosphorane gave after photo-isomerization with a catalytic amount of iodine the diene aldehyde 42 representing the bottom portion of the tetraene epoxide 10.

The top portion was the same as for the syntheses of the lipoxin B isomers therefore treatment of the phosphonium salt 30 in THF with 0.95 equivalent of lithium hexamethyldisilazide at -100° for one min. then addition of the aldehyde 43 (1.0 equivalent) gave after warming to -40° for 2 hours the desired protected tetraene epoxide and its all trans isomer in equal amounts.

All initial attempts to remove the silyl protecting group with nBu$_4$NF met with dismal failure, resulting consistently in the complete loss of recognizable material. This was reasoned to be at least partially due to the alkoxide generated upon removal of the silyl, therefore it was decided to add a small amount of acetic acid to the reaction mixture to quench the alkoxide, despite the exquisite acid lability of the tetraene epoxide moiety.[13] This strategy proved successful yielding a mixture of the alcohol ester and the δ lactone.

Sponification of the ester and/or lactone with methanolic sodium hydroxide followed by aqueous hydrolysis of the tetraene epoxide 10 at various pH's consistantly gave the six lipoxin isomers, (Figure 11), complementary to the six generated by aqueous hydrolysis of the 5,6 tetraene epoxide 9. The ratios of the various isomers varied only slightly with pH. While the 8-cis-lipoxin B isomer 8 is produced by aqueous hydrolysis of the tetraene epoxide 10 its C-14 epimer 5 is also produced in comparable amounts. However in biologically generated mixtures of lipoxins both cis isomers have not been found in a similar ratio. Therefore, with the lack of the 14S diasteriomer in samples containing 5S,14R,15S-8-cis-lipoxin B 6 it is unlikely that this isomer arises principally from the aqueous hydrolysis of the tetraenes epoxide 10. This isomer may however arise from lipoxygenation of 5,15 diHPETE or enzymatic hydrolysis of a tetraene epoxide.

However if this isomer 8 arises from conjugate enzymatic hydrolysis of the tetraene epoxide 9 the mechanism must be different than that postulated for the conversion of LTA$_4$ to LTB$_4$ which involves addition of water from the face of the molecule opposite to the epoxide to give the 12R stereocenter and the 6Z,8E,10E triene system. Applying these criteria to epoxide 9 to yield a 6,10,12E,8Z tetrane system we see that the newly generated center would be S again due to the difference in priorities caused by having a CH(OH) rather than a CH$_2$ adjacent to this center. It is interesting to note that the generation of 5S,14S,15S-8-cis lipoxin B 7 and 5S,6R,14S-11-cis lipoxin A 2 would involve addition of water from the same face. The lipoxin A isomer 1 and the lipoxin B isomer 8 are also related in this manner.

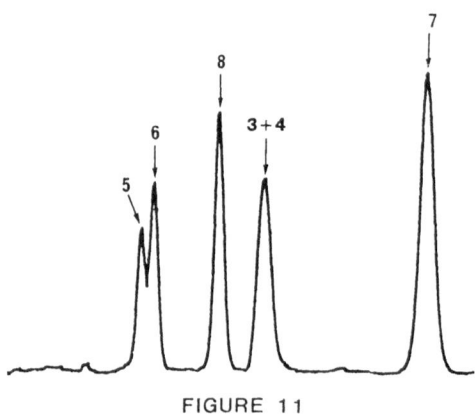

FIGURE 11

In conclusion the syntheses of eight bio-rational lipoxin isomers of unambiguous relative, absolute and double bond stereochemistry was accomplished by several routes. The availability of these synthetically pure standards has allowed for the identification of seven lipoxin isomers from the incubation of 15-HPETE with human leucocytes. Two possible biogenic precursors to the lipoxins were also prepared. These tetraene epoxides 9 and 10 have facilitated the study of lipoxin biosynthesis. The above results illustrate the key role synthetic organic chemistry plays in the study of biologically derived compounds.

REFERENCES

1. Serhan, C.N.; Hamberg, M.; Samuelsson, B. Biochem. Biophys. Res. Commun. 1984, 118, 943-949.

2. Serhan, C.N.; Hamberg, M.; Samuelsson, B. Proc. Natl. Acad. Sci. U.S.A. 1984, 81, 5335-5339.

3. Samuelsson, B. Adv. Prost. Throm. Leuko. Res. 1985, 15, 1-9.

4. Ramstedt, U.; Ng, J.; Wigzell, H.; Serhan, C.N. and Samuelsson, B. J. Immunol. 1985, 135, 3434-3438.

5. Hanson, A.; Serhan, C.N. ; Haeggstrom, J.; Ingelman-Sandberg, M. and Samuelsson, B. Biochem. Biophys. Res. Commun. 1986, 134.

6. Adams, J.; Fitzsimmons, B.J. and Rokach, J. Tetrahedron Lett. 1984, 25, 4713-4716.

7. Adams, J.; Fitzsimmons, B.J.; Girard, Y.; Leblanc, Y.; Evans, J.F. and Rokach, J. J. Am. Chem. Soc. 1985, 107, 464-469.

8. Leblanc, Y.; Fitzsimmons, B.; Adams, J. and Rokach, J. Tetrahedron Lett. 1985, 26, 1399-1402.

9. Fitzsimmons, B.J.; Adams, J.; Evans, J.F.; Leblanc, Y. and Rokach, J. J. Biol. Chem. 1985, 260, 13008-13012.

10. Fitzsimmons, B.J. and Rokach, J. Tetrahedron Lett. 1985, 26, 3939-3942.

11. Serhan, C.N.; Hamberg, M.; Samuelsson, B.; Morris, J. and Wishka D.G. Proc. Natl. Acad. Sci. 1986, 83, 1983-1987.

12. Rokach, J.; Fitzsimmons, B.J.; Leblanc, Y.; Ueda, N. and Yamamoto, S. Adv. Prost. Throm. Leuko. Res. in press. Presented at the 6th International Conference on Prostaglandins and Related Compounds, Florence, Italy, June 3-6, 1986.

13. The half life of this compound in PH 7.4 buffer is approximated at less than two minutes.

14. Corey, E.J.; Mehrotra, M.M. and Su, W.-g. Tetrahedron Lett. 1985, 26, 1919-1922.

8

COMPUTED CONFORMATIONAL ANALYSIS OF LIPOXINS

AND THEIR IONIC COMPLEXES

Robert Brasseur, Charles N. Serhan and Michel Deleers

Macromolecules at Interfaces, CP 206/2, Brussels
Free University, B-1050 Brussels, Belgium
Hematology Division of the Brigham and Women's
Hospital and Harvard Medical School
75 Francis Street, Boston, Ma   02115
Research Center, UCB Pharmaceutical Sector, B-1420
Braine l'Alleud, Belgium

ABSTRACT

The possible molecular conformations of four structurally
and biologically different lipoxins derivatives were predicted
by a systematic structure tree theoretical analysis. This
method takes into account the London-Van der Waals energy of
interaction, the electrostatic interaction, the rotation
energy of the torsional angles and the energy of transfer
through a possible lipid-water interface. Finally, the
conformers derived from the structure tree and with a high
probability of existence were submitted to the energy minimization procedure. The most probable conformers of lipoxin
A: 5S,6R,15S-trihydroxy-7,9,13 trans-11 cis -eicosatetraenoic
acid (LXA); 11 trans lipoxin A: 5S,6R,15S-trihydroxy-7,9,11,
13 trans -eicosatetraenoic acid (11t-LXA); lipoxin B: 5S,14R
15S-trihydroxy-6,10,12 trans-8 cis -eicosatetraenoic acid
(LXB) and 8 trans lipoxin B: 5S,14R,15S-trihydroxy- 6,8,10,
12 trans -eicosatetraenoic acid (8t-LXB) in their isolated
form or when forming complexes with one calcium ion are
presented.

The four isolated compounds lead to vastly different
conformations. Lipoxin A can form the most globular conformer
while lipoxin B seems to be slightly more extended.   The all
trans isomer of lipoxin B forms an extended conformer and
11 trans lipoxin A gives a fully extended molecule. Complexes
of a pair of these compounds with one calcium ion were shown
to lead to vastly different conformations. Both $(LXA)_2Ca$ and
$(LXB)_2Ca$ form crumpled or extended structure, the LXA molecules being more wrapped around Ca^{2+} than LXB molecules.
The $(11t-LXA)_2Ca$ and $(8t-LXB)_2Ca$ complexes present a high
probability of extended conformations. Our description
merely shows that the peculiar stereochemistry of these

molecules lead to equilibria between conformers or to very
static conformers, the flexibility and rigidity of which being
probably relevant in view of their different biological
activities.

INTRODUCTION

The formation of a number of oxygenated derivatives of
arachidonic acid in leukocytes appears to be involved in
inflammatory and immunological processes (Samuelsson, 1983,
Needleman et al.,1986). Recently, the stereochemistries of
lipoxins that are formed by the interaction between the 5-
and 15-lipoxygenase pathways in A23187-activated leukocytes
incubated with 15 HPETE (or 15 HETE)(Serhan et al., 1984a)
were fully determined (Serhan et al., 1984b, 1986a, 1986b).
Because these compounds display highly stereospecific biological activities (Serhan et al., 1985; Ramstedt et al., 1985,
1987; Hansson et al., 1986; Dahlen et al., 1987) , LXA and
LXB being always more active than their all trans conformers,
we have attempted in the present study to define the conformation of these molecules in their isolated carboxylic acid
form or when forming possible complexes with calcium (Deleers
et al., 1983, 1985; Brasseur and Deleers, 1983 , 1984).

The great agreement between the prediction method and
experimental observation has put forward the analysis of
conformations for molecular recognition (Rebek, 1987).
Indeed, it appears to be a powerfull tool to visualize
molecules and to obtain informations on interactions between
compounds (Brasseur et al., 1983a, 1983b), on structure
activity relationship of pharmacological drugs or biological
compounds (Ralston et al., 1974; Basak et al., 1982; Duax et
al., 1983; Deleers et al., 1983a, 1983b, 1983c), but also on
the mode of microorganization of amphiphilic molecules in
the vicinity of membranes (Brasseur et al., 1982, 1984, 1986;
Brasseur 1986).

METHODS

Our theoretical prediction (molecular structure analysis)
of each compounds and possible Ca^{2+} complexes is based on the
evaluation of several conformational parameters. The total
conformational energy is calculated as the sum of the following terms:
1) The London-Van der Waals energy of interaction between all
pairs of non mutually bonded atoms. The Buckingam's pairwise
atom-atom interaction functions have been used:

$$E_{vdw} = \sum_{ij} [A_{ij} \exp(-B_{ij}r_{ij}) - C_{ij}r_{ij}^{-6}]$$

where i, j = 1,2,... are nonbonded atoms, r_{ij} their distances
from each other, and A_{ij}, B_{ij} and C_{ij} coefficients assigned
to atom pairs. . The values of these coefficients have been
reported by Liquori and co-workers (1968, 1969). They have
been applied with success to conformational analysis of molecular crystals, proteins, polypeptides and amphiphilic molecules. In order to compensate for the low values of the
function E_{vdw} at small r_{ij}, we have imposed an arbitrary

94

cut-off value of $E_{vdw} = 100$ Kcal/mol at $r_{ij} \leqslant 1$ Å.
2) The generalized Keesom-Van der Waals interaction or electrostatic interaction between atomic charges:

$$E_{cb} = 332 \sum_{ij} \frac{e_i \, e_j}{r_{ij} \, \varepsilon_{ij}}$$

where e_i and e_j are expressed in electron charge units and r_{ij} in angströms. ε_{ij} is the dielectric constant which is subject to variations.. The values of atomic point charge are identical to the values used for polypeptides (Ralston and DeCoen 1974, DeCoen and Ralston 1977).
3) The potential energy of rotation of torsional angles: This rotation around the C-C or C-O bonds was calculated by

$$E_{tor} = \frac{U_{ij}}{2} \cdot (1 + \cos \Phi_{ij})$$

where U_{ij} corresponds to the energy barrier in the eclipsed conformation during the rotation of the angle, and Φ_{ij} is the torsional angle. U_{ij} is equal to 2.8 Kcal/mol for the C-C bond and 1.8 Kcal/mol for the C-O bond.

The conformational energy is calculated for a large number of conformations in a systematic analysis structure tree (Brasseur and Deleers 1983, 1984), where six changes of 60° each were successively imposed to the n torsional angles choosen, yielding 6^n conformers in each branch of the structure tree. The conformational energy was calculated for each of these conformers and the most probable configurations were taken as those yielding the lowest energy by the following equation:

$$P = \exp(-E_j/RT) \; / \sum_i^n \exp(-E_i/RT)$$

with $T = 25°$, and E_j E_i corresponding respectively to the internal energy of the conformer under consideration and the energies of all conformers generated. The effect of entropy was considered at this stage of calculation process as negligible and, hence, the selection of conformers was based on their energy rather than free energy. The conformations derived from the systematic study and yielding a low internal energy were submitted to a second analysis, using a simplex minimization procedure (Nelder and Mead, 1965) in order to further reduce their total energy. All torsional angles were examined in such an analysis with a precision of 10° on each conformational angle. The values used for the valence angles and boundary lengths are those currently used in conformational analysis (Liquori, 1969; Hopfinger, 1973; Ralston and DeCoen, 1974). The systematic and minimization procedures were made in a medium of intermediate dielectric constant representative of the vicinity of a membrane-water interface. In a second minimization procedure, the hydrophilic and hydrophobic gravity centers of selected conformers were established at each step of this procedure and take into account the transfer energy of each part of the molecule (CH3-, -CH2-, -CH=, -OH, =C=O, -Ca-) relative to a possible interface. This allows to calculate the hydrophobic-hydrophilic balance and the distance between the hydrophobic and hydrophilic gravity centers (Brasseur et al., 1986; Brasseur, 1986).

Calculations were made either on a CDC Cyber 170 computer coupled to a calcomp 1051 drawning table (PLUTO program: Motherwell and Clegg, 1975), or on an Olivetti M24 (M28)

computer equipped with a 8087 arithmetic coprocessor with the
aid of the PC-MSA$_{(r)}$ program (molecular structure analysis),
the PC-MGM$_{(r)}$ program (molecular graphic manipulation) and the
PC-TAMMO program (theoretical analysis of membrane molecular
organization : Brasseur, 1986). These programs also run on
either IBM XT, AT386, AT&T or every IBM PC compatible of at
least 640 Kram + 2 Mram disk equipped with hard disk and
arithmetic coprocessor.

RESULTS

     The molecular structure and the conformation of the four
lipoxins taken as our initial conformers are those illustrated
elsewhere (Serhan etal., 1986$_a$, 1986$_b$), except that LXA and
LXB are quasi-extended at the beginning of the computing
procedure. Each molecule has 11 to 13 important rotational
angles and if these angles were affected by systematic 60°
changes, more than 362.10^6 conformations could be designated.
To avoid this large number of conformations, another procedure
was used. Systematic analysis was carried out in a stepwise
manner on different parts of the molecules, generally by
sequences of 5 rotational bounds (7776 conformations) and
gives a structure tree for each molecule.. Figure 1a,b,c,d
summarizes the most probable configurations of LXA,
11t-LXA, LXB and 8t-LXB shown in stereoscopic views.

     When complexes are formed with calcium, the same procedure
was carried out on half of the complexes taking into account
the bound between Ca^{2+} and O-C-, and gives one structure tree
for each complex.. There is increasing evidences that calcium
complexes with two molecules are symmetrical or quasi-symme-
trical (Brasseur et al., 1982; Deleers et al., 1983$_a$, 1983$_b$,
1983$_c$). Table 1 summarizes the probabilities of the most
probable configurations of LXA-Ca and LXB-Ca complexes (selection based on a Boltzman statistical weight of all configurations) with their probability of existence in a structure
tree obtained after three successive systematical analysis.

Table 1. Structure tree of LXA and LXB complexes with
          Ca^{2+} after systematic theoretical analysis.

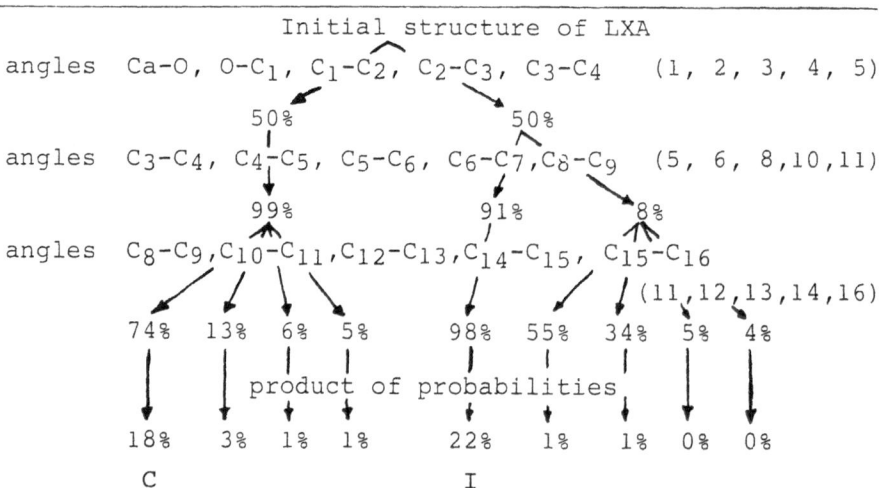

96

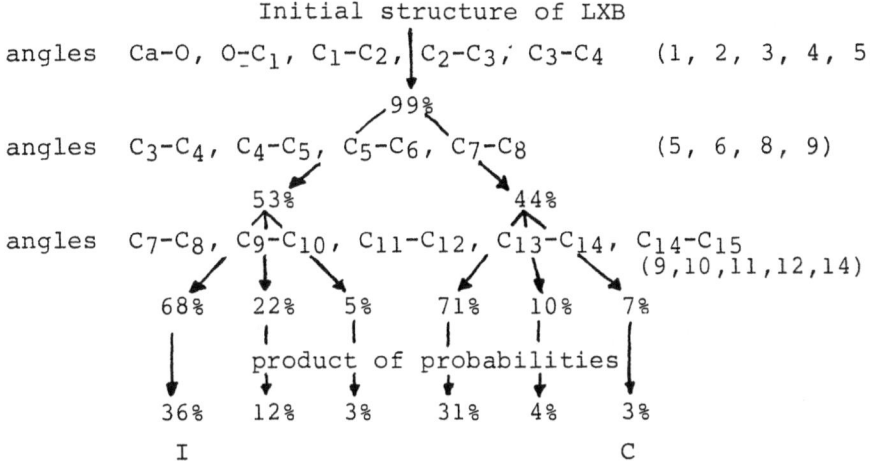

Initial structure of LXB

angles Ca-O, O-C$_1$, C$_1$-C$_2$, C$_2$-C$_3$, C$_3$-C$_4$ (1, 2, 3, 4, 5)

                        99%

angles C$_3$-C$_4$, C$_4$-C$_5$, C$_5$-C$_6$, C$_7$-C$_8$ (5, 6, 8, 9)

            53%   44%

angles C$_7$-C$_8$, C$_9$-C$_{10}$, C$_{11}$-C$_{12}$, C$_{13}$-C$_{14}$, C$_{14}$-C$_{15}$
                                    (9,10,11,12,14)

      68%   22%   5%   71%   10%   7%

                product of probabilities

      36%   12%   3%   31%   4%   3%
       I   C

In table 1, I and C denote respectively extended (or interfacial) and cryptic configurations. It is obvious that an equilibrium between the two form may exist with lipoxin A while the extended structure is favoured in the case of lipoxin B. It is also evident that the LXA-Ca complex adopts two or three configurations whereas the LXB-Ca complex seems to have more possibilities. Table 2 indicates the orientation of torsional angles after the minimization procedure bearing this time on all angles of half of the complex formed with Ca^{2+} and two molecules of LXA, 11t-LXA, LXB or 8t-LXB.

Table 2. Values of the torsional angles after minimization procedure.

	LXA		11t-LXA		LXB		8t-LXB
angles	I	C	I	angles	I	C	I
Ca-O	177	63	163	Ca-O	180	177	181
O-C$_1$	3	340	30	O-C$_1$	359	358	3
C$_1$-C$_2$	351	1	341	C$_1$-C$_2$	355	355	359
C$_2$-C$_3$	285	294	296	C$_2$-C$_3$	294	292	288
C$_3$-C$_4$	214	205	222	C$_3$-C$_4$	213	215	216
C$_4$-C$_5$	173	185	149	C$_4$-C$_5$	179	170	180
C$_5$-OH	176	174	192	C$_5$-OH	163	164	159
C$_5$-C$_6$	174	62	304	C$_5$-C$_6$	174	177	177
C$_6$-OH	165	173	211	C$_7$-C$_8$	174	176	356
C$_6$-C$_7$	173	52	139	C$_9$-C$_{10}$	178	178	175
C$_8$-C$_9$	182	164	182	C$_{11}$-C$_{12}$	0	181	176
C$_{10}$-C$_{11}$	11	349	179	C$_{13}$-C$_{14}$	178	235	111
C$_{12}$-C$_{13}$	192	174	174	C$_{14}$-OH	180	179	175
C$_{14}$-C$_{15}$	197	181	196	C$_{14}$-C$_{15}$	296	62	298
C$_{15}$-OH	186	188	195	C$_{15}$-OH	179	179	185
C$_{15}$-C$_{16}$	108	122	90	C$_{15}$-C$_{16}$	174	174	172
C$_{16}$-C$_{17}$	187	178	186	C$_{16}$-C$_{17}$	178	177	178
C$_{17}$-C$_{18}$	178	182	191	C$_{17}$-C$_{18}$	177	180	178
C$_{18}$-C$_{19}$	177	168	185	C$_{18}$-C$_{19}$	174	177	179
distance	4.7	0.8	5.2		3.7	1.4	4.8
figures		2$_a$	2$_b$		3$_a$	3$_b$	

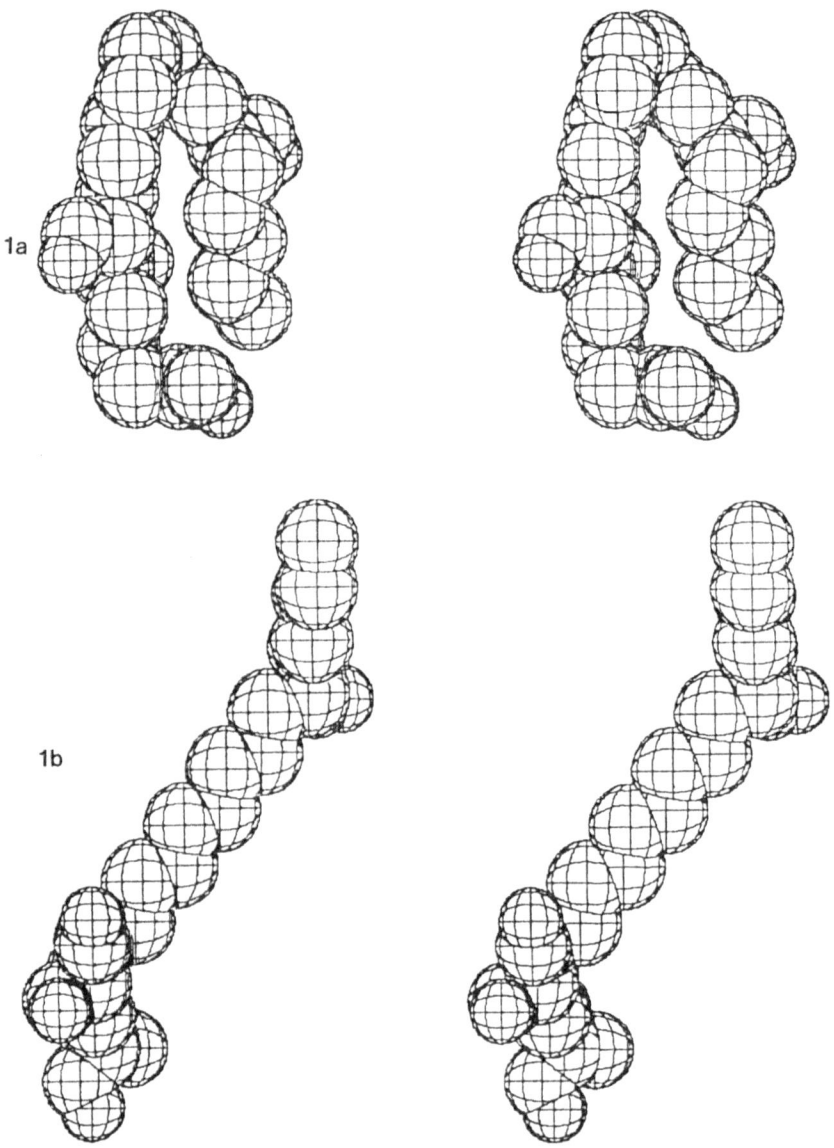

Figs 1_a & 1_b. Stereoscopic views of the most probable conformations of lipoxin A and 11 trans-lipoxin A in space filling- cross hatching representation. Solely the hydrogens of the 3 hydroxy functions are drawn.

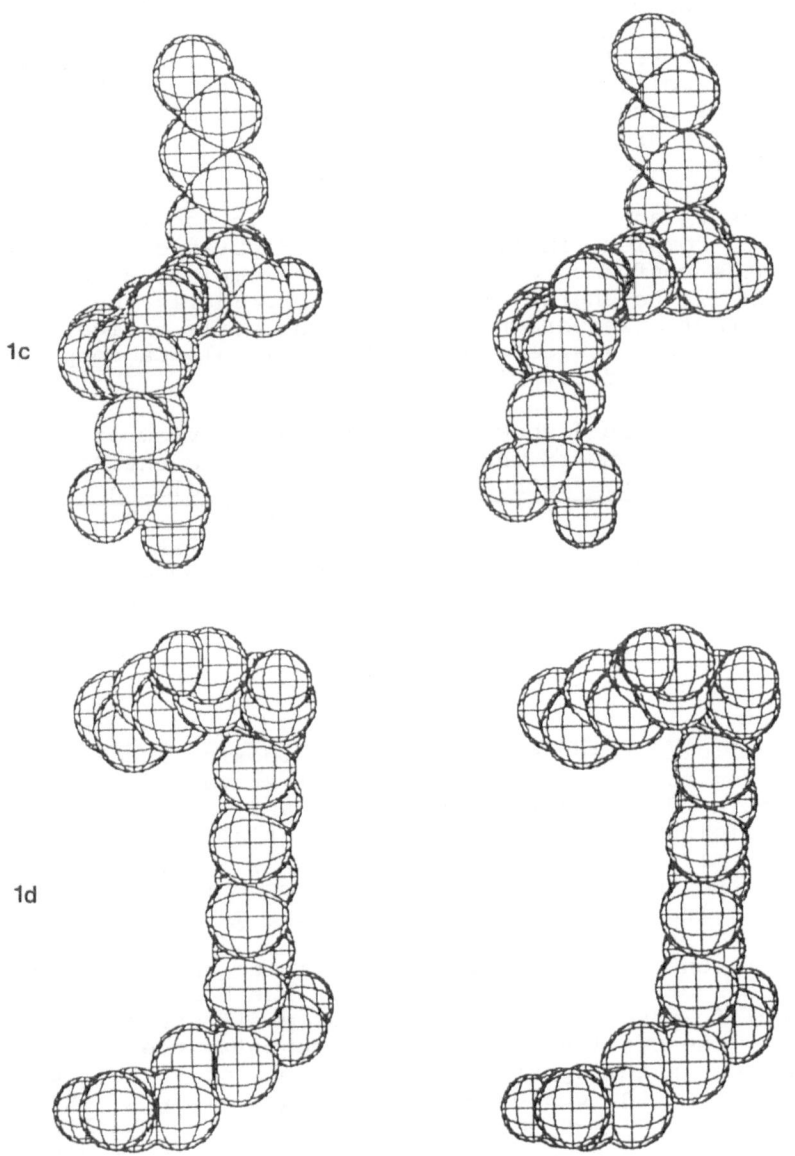

1c

1d

Figs 1c & 1d. Stereoscopic views of the most probable conformations of lipoxin B and 8 trans-lipoxin B in space filling- cross hatching representation. Solely the hydrogens of the 3 hydroxy functions are drawn.

It is clear that LXA(I) differs from LXA(C) by six angles by at least 15° whereas LXB(I) differs from LXB(C) by only 3 angles. This may account in part at least for the more globular form (or extended form) as compared to LXB. This is also shown in table 2 where the variation in the distance between the hydrophobic and hydrophilic gravity centers is more pronounced in LXA than in LXB. The hydrophobic transfer energy is 83.28 Kcal/mol and the hydrophilic transfer energy is 51.84 Kcal/mol (Tanford, 1973), i.e., an hydrophobic-hydrophilic balance that corresponds to water soluble molecules which may easily cross a membrane or interact with it (Brasseur et al., 1986).

Figures 2_a and 2_b show the stereoscopic conformations of $(LXA)_2$-Ca complex in its cryptic configuration and the most probable conformer of $(11t-LXA)_2$-Ca complex after application of the energy minimization procedure, the all-trans conformer being drawn at 3/4 of the scale. Figures 3_a and 3_b show the stereoscopic views of $(LXB)_2$-Ca complex in the extended and cryptic configurations.

Finally, figure 4 shows the stereoscopic view of $(14S-8t-LXB)_2$-Ca complex when calculated on the entire complex by the structure tree method in a procedure yielding 1,353,024 different structures. This figure clearly shows that the complex is symmetrical or quasi symmetrical after a calculation procedure on 27 torsional angles and merely demonstrates the usefulness of the actual method bearing on only half of the complex.

DISCUSSION

The cellular responses induced by lipoxins appear to be distinct from those evoked by other derivatives of the arachidonic acid metabolism. Lipoxin A and lipoxin B have different biological activities (Serhan et al., 1986_a, 1986_b; Ramstedt et al., 1985, 1987; Dahlen et al., 1987; Hansson et al., 1985) and their responses are stereospecific in the sense that the all-trans isomers are always less potent.

The present work aims at characterizing the most probable conformations of 4 lipoxins selected on the basis of their stereospecific responses and their vastly different biological potencies. the results of our study emphasize the existence of difference in the spatial configurations of these lipoxins. It is clear from the conformations in figure $1_{a,b,c,d}$ that a sequence of folding may be written as follows : LXA > LXB > 8t-LXB > 11t-LXA. From figures $2_{a,b}$ and $3_{a,b}$ it is also obvious that LXA and LXB may give extended structures or cryptic structures, the wrapping of lipoxin A around Ca^{2+} being more pronounced than that of its position isomer LXB. This is also clear in view of the distance between the hydrophobic and hydrophilic gravity centers. The globular configuration and the low distance between these gravity centers may be important. Indeed, we have already shown that this distance may account either for the calciphoretic properties of leukotriene B_4 (Serhan et al.,1982; Brasseur and Deleers, 1984) or for the activity of other ionophores (Brasseur et al., 1982, 1984, 1986; Deleers et al., 1983_c).

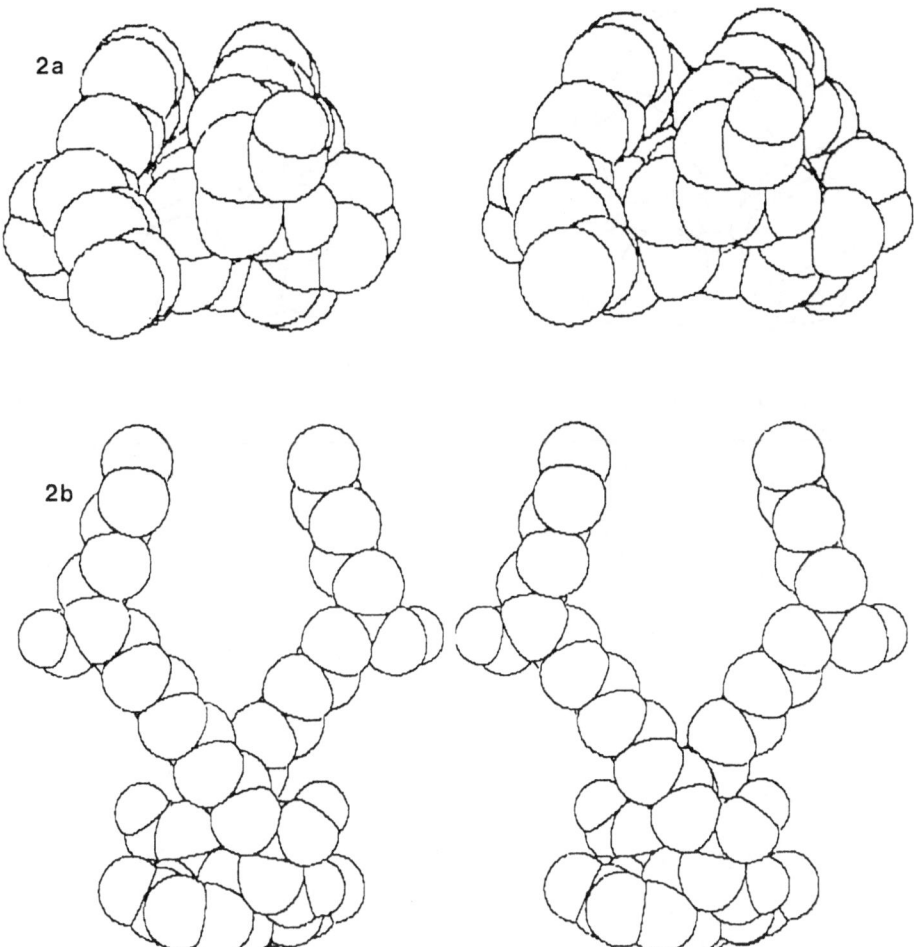

Figs 2$_a$ & 2$_b$. Stereoscopic representations (space filling)
of the (LXA)$_2$Ca complex (top) in its folded configuration
and the most probable conformer of (11t-LXA)$_2$Ca complex
(bottom)(see table 2). The hydrogens bonded to C are not
drawn, only the H of the hydroxy groups are shown. The
all-trans conformer complex (bottom) is drawn at 3/4 of
the LXA scale (top).

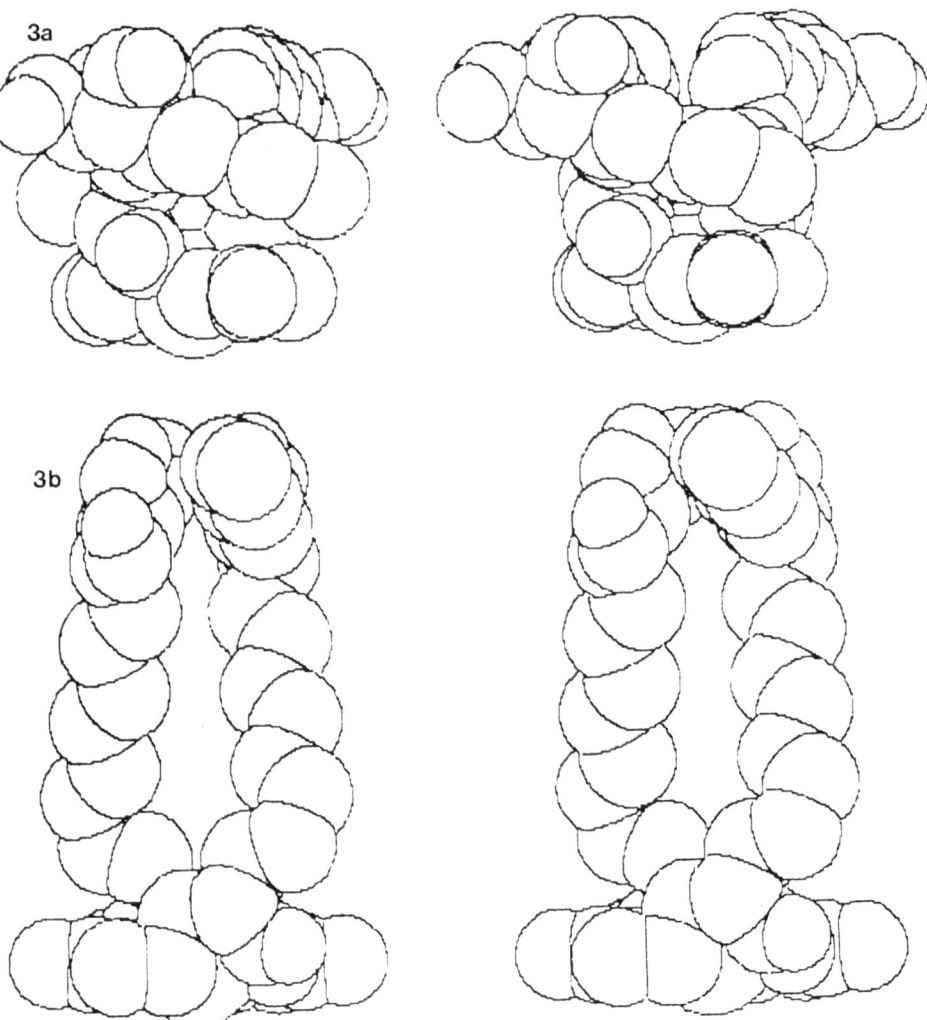

Figs 3_a & 3_b: Stereoscopic representations (space filling) of the $(LXB)_2Ca$ complexes in their cryptic (top) and extended (interfacial) configurations (bottom)(see table 2). The hydrogens bonded to C are not drawn, only the hydrogens of the hydroxy groups are shown. The scale is identical to the scale of fig 2_a.

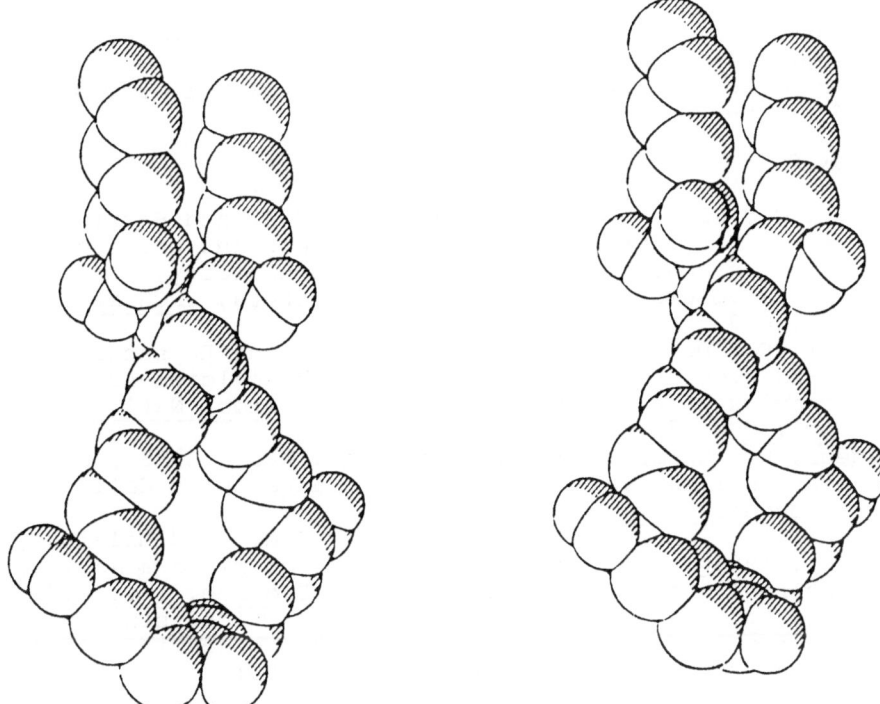

Figure 4. Stereoscopic view of the $(14S,-8t-LXB)_2-Ca$ complex in shaded-space filling representation and obtained after systematic analysis bearing on all torsional angles and minimization procedure (CDC cyber 170 computer).

The results of our molecular structure analysis may help in understanding differences in the biological potency of the four lipoxins studied if the assumption is made that the flexibility and rigidity is relevant of the expression of their biological properties. For instance, the biological activity of LXA could be explained by the equilibrium between the globular and extended form, a mechanism which demonstrates a great flexibility and hence, an easier interaction with a probable receptor. In this view, the possibility to calculate with our method, the conformation of molecules in a lipid-water interface (transfer energies) separating two media of distinct dielectric constant (Brasseur et al., 1984, 1986; Brasseur,1986) appears to be promising. The lesser activity of LXB may then be explained by the equilibrium between the two forms that favors the extended conformation.

Taken together, the results of these spatial conformations performed with the same procedure on the four isomeric compounds indicate that they each lead to vastly different structures, either in the absence or presence of calcium. It appears that the differences in spatial configurations may account, in part at least, for the different biological activities of these molecules (Serhan et al., 1986[b]; Dahlen et al., 1987) and may provide the basis for their cellular recognition. The availability of calcium plays a critical role in both the biosynthesis and action of eicosanoids

(Samuelsson, 1983). The present results may also provide
a conformational basis for the interactions between Ca^{2+}
and eicosanoid substrate-enzyme complex, or for the interactions between receptors or proteins involved in signal
transduction-eicosanoid interactions.

REFERENCES

Basak, S.C., Gierchen, D.P., Magnusson, V.R. and Harris, D.K.,
    1982, Structure activity relationship and pharmacokine-
    tics: a comparative study of hydrophobicity, Van der
    Waals volume and topological parameters, IRCS Med. Sci.
    10: 619.
Brasseur, R., Deleers, M., Malaisse, W.J. and Ruysschaert, J.M.
    1982, Conformational analysis of the calcium-A23187
    complex at a lipid-water interface, Proc. Natl. Acad.
    Sci. USA, 79: 2895.
Brasseur, R., Deleers, M., and Malaisse, W.J., 1983[a], Confor-
    mational analysis of the calcium antagonist Gallopamil,
    Biochem. Pharmacol.,32: 437.
Brasseur, R., Deleers, M., Ruysschaert, J.M., and Malaisse, W.
    J., 1983[b], Conformational analysis of mixed monolayers of
    phorbol esters and phospholipids, Biochem. Int. 7:71.
Brasseur, R., and Deleers, M., 1983, Theoretical study on
    conformation related activity of hypoglycemic sulfonyl-
    ureas, Pharmacol. Res. Commun., 15: 901.
Brasseur, R., Deleers, M., and Ruysschaert, J.M., 1984,
    Sequence of ionophore conformation changes induced by a
    simulated membrane-water interface, Biosc. Rpts.,4: 651.
Brasseur, R., and Deleers, M., 1984, Conformational analysis
    of 6-cis and 6-trans leukotrienes-B4-Ca^{2+} complexes,
    Proc. Natl. Acad. Sci. USA, 81: 3370.
Brasseur, R., Deleers, M., and Ruysschaert, J.M., 1986,
    Localization and mode of organization of amphiphilic
    molecules at a lipid-water interface: a predictional
    approach, J. Colloid Interface Sci., 114: 277.
Brasseur, R., 1986, Theoretical analysis of membrane molecular
    organization, J. Molec. Graphics, 4: 117.
Brasseur, R., Deleers, M., Ruysschaert, J.M., Samuelsson, B.,
    and Serhan, C.N., 1987, Conformational analysis of
    lipoxin A, lipoxin B and their isomers, Submitted.
Dahlen, S.E., Rand, J., Serhan, C.N., Björk, J., and
    Samuelsson, B., 1987, Biological activities of lipoxin A
    includes lung strip contraction and dilation of arterio-
    les in vivo, Acta Physiol. Scand., submitted.
DeCoen, J.L., and Ralston, E., 1977, Theoretical conformatio-
    nal analysis of Asn1,Val5 angiotensin, Biopol.,73 : 38.
Deleers, M., Brasseur, R., Ruysschaert, J.M., and Malaisse, W.
    J.,1983[a], Conformational analysis of phorbol esters at a
    simulated membrane water interface, Biophys. Chem. ,
    17: 313.
Deleers, M., Brasseur, R., and Malaisse, W.J., 1983[b] ,
    Stoichiometry of calcium binding by hypoglycemic sulfo-
    nyl ureas, Res. Commun. Chem. Pathol. Pharmacol., 42:
    181.
Deleers, M., Brasseur, R., and Malaisse, W.J., 1983[c], Calcium
    transport by a beta-diketone in model membranes, Chem.
    Phys. Lipids, 33: 11.

Deleers, M., Grognet, P., and Brasseur, R., 1985, Structural
    considerations for calcium ionophoresis by prostaglandins,
    Biochem. Pharmacol., 34: 3831.
Duax, W.L., Smith, G.D., Griffin, J.F. and Portoghese,P., 1983,
    Methadone conformation and opioid activity, Science,220:
    417.
Hansson, A., Serhan, C.N., Haeggström, J., Ingelman-Sundberg,
    M., and Samuelsson, B., 1986, Activation of protein
    kinase C by lipoxin A and other eicosanoids: intracellu-
    lar action of oxygenation products of arachidonic acid,
    Biochem. Biophys. Res. Commun., 134 : 1215.
Hopfinger, A.J., 1973,"Conformational properties of macromole-
    cules", Academic Press, New York.
Liquori, A.M., Giglio, E., and Mozzarella, L., 1968, Van der
    Waals interactions and the packing of molecular crystals,
    Nuovo Cimento, 55 : 475.
Liquori, A.M., 1969, The stereochemical code and the logic of
    a protein molecule, Quat. Rev. Biophys., 2 : 65.
Motherwell, S., and Clegg, W., 1978, "Pluto program", Univer-
    sity of Cambridge, Cambridge University Press, London.
Needleman, P., Turk, J., Jakschik, B.A., Morrison, A.R., and
    Lefkowith, J.B., 1986, Arachidonic acid metabolism,
    Ann. Rev. Biochem., 55 : 69.
Nelder, J.A., and Mead, R., 1965, A simplex method for
    function minimization, Computer J., 7: 308.
Ralston, E., and DeCoen, J.L., 1974, Folding of polypeptide
    chains induced by the amino side chains, J. Mol. Biol.,
    83 : 393.
Ralston, E., DeCoen, J.L., and Walter, R., 1974, Tertiary
    structure of H-Pro-Leu-Gly-NH$_2$, the factor that inhibits
    release of melanocyte stimulating hormone, derived by
    conformational energy calculations, Proc. Natl. Acad.
    Sci. USA, 71 :1142.
Ramstedt, U., Ng, J., Wigzell, H., Serhan, C.N., Samuelsson,
    B., 1985, Action of novel eicosanoids lipoxin A and B
    on human natural killer cell cytotoxicity: effects on
    intracellular cAMP and target cell binding,
    J. Immunol., 135 : 3434.
Ramstedt, U., Serhan, C.N., Nicolaou, K.C., Weber, S.E.,
    Wigzell, H., and Samuelsson, B., 1987, Lipoxin A induced
    inhibition of natural killer cells: studies on stereo-
    specificity and mode of action, J. Immunol., 138 : 266.
Rebek, J., 1987, Model studies in molecular recognition,
    Science, 235 : 1478.
Samuelsson, B., 1983, Leukotrienes: mediators of immediate
    hypersensitivity reaction and inflammation, Science,
    220 : 568.
Serhan, C.N., Fridovitch, J., Goetzl, E., Dunham, P.B. and
    Weissmann, G., 1982, Leukotriene B4 and phosphatidic
    acid are calcium ionophores, J. Biol. Chem., 257: 4746.
Serhan, C.N., Hamberg, M., and Samuelsson, B., 1984$_a$,
    Trihydroxytetraenes: a novel series of compounds formed
    from arachidonic acid in human leukocytes, Biochem.
    Biophys. Res. Commun., 118 : 943.
Serhan, C.N., Hamberg, M., and Samuelsson, B., 1984$_b$, Lipoxins:
    novel series of biologically active compounds formed
    from arachidonic acid in human leukocytes, Proc. Natl.
    Acad. Sci. USA, 81 : 5335.

Serhan, C.N., Hamberg, M., and Samuelsson, B., 1985, Lipoxins,
a novel series of biologically compounds, in: "Prosta-
glandins, leukotrienes and lipoxins" Bailey Ed. New York.
Serhan, C.N., Hamberg, M., Samuelsson, B., Morris, J., and
Wishka, D.J., 1986_a, On the stereochemistry and biosyn-
thesis of lipoxin B, Proc. Natl. Acad. Sci. USA, 83 :
1983.
Serhan, C.N., Nicolaou, K.C., Webber, S.E., Veale, C.A.,
Dahlen, S.E., Puutsinen, T.J., and Samuelsson, B., 1986_b,
Lipoxin A: stereochemistry and biosynthesis, J. Biol.
Chem., 261 : 16340.
Tanford, C., 1973, "The hydrophobic effects. Formation of
micelles and biological membranes ", John Wiley and sons,
New York.

9

ACTIONS OF LIPOXIN A_4 AND RELATED COMPOUNDS

IN SMOOTH MUSCLE PREPARATIONS AND

ON THE MICROCIRCULATION IN VIVO

Sven-Erik Dahlén[a], Lilian Franzén[a],
Johan Raud[a], Charles N. Serhan[b,c],
Pär Westlund[b], Eva Wikström[a],
Thure Björck[a], Hisao Matsuda[a],
Stephen E. Webber[d], Chris A. Veale[d],
Tapio Puustinen[b], Jesper Haeggström[b],
K.C. Nicolaou[d], and Bengt Samuelsson[b]

[a]Department of Physiology
Karolinska Institutet, and
the National Institute of
Environmental Medicine
Stockholm, Sweden

[b]Department of Physiological Chemistry
Karolinska Institutet
Stockholm, Sweden

[c]Hematology Division, Brigham and Women's
Hospital and Harvard Medical School
Boston, Mass.,USA

[d]Department of Chemistry, University of
Pennsylvania, Philadelphia, USA

INTRODUCTION

The present chapter summarizes our findings with lipoxins (LX) in spasmogenic assays and in the intact microvasculature of the hamster cheek pouch. The initial observations[1,2] were made in experiments using lipoxins isolated from human leukocytes. With the aid of synthetic compounds[3], it has been possible to further explore the pharmacodynamics and structure activity relationships for lipoxins in smooth muscle preparations.

SPECTRUM OF ACTIVITIES

Microcirculation

In the hamster cheek pouch, prepared for intravital microscopy of the terminal vascular network [4], topical administration of LXA$_4$ (5S,6R,15S-trihydroxy-7,9,13-trans-11-cis-eicosatetraenoic acid, 1 μM) induced a pronounced arteriolar dilation, but did not change venular diameters [1]. The dilation of arterioles started to develop within 30 seconds, gradually increased to peak at the end of the application period (3 min), and readily disappeared thereafter.

On the other hand, LXA$_4$ (tested up to 3 μM) did not affect FITC-dextran extravasation, used as a marker for microvascular permeability to plasma proteins, nor was leukocyte adherence to the endothelium of small venules stimulated by LXA$_4$. It is unlikely that the vasodilator property of LXA$_4$ masked an influence on circulating leukocytes, because vasodilation induced by prostaglandin(PG) E$_2$ markedly enhances leukocyte activation elicited with leukotriene(LT) B$_4$ in the hamster cheek pouch [5].

Evidently, the microvascular activity of LXA$_4$ in the hamster cheek pouch was different from the profiles of activities known for the leukotrienes (Fig.1.)[6]. Finally, the microvascular responses obtained with LXA$_4$ isolated from leukocytes has recently been corroborated with stereochemically defined synthetic material.

Smooth muscle preparations

Using biologically isolated material, LXA$_4$ was found to contract the guinea pig lung strip but not strips of guinea pig ileum or trachea [2]. This pattern of spasmogenic activity is thus distinct from the effects of either LTB$_4$ or LTC$_4$ in these three assays(Fig.2). In addition, there is no known cyclooxygenase product which has this profile of contractile activity [7].

The contractile potency of LXA$_4$ in the guinea pig lung strip was indicated by half maximal contractions obtained at

MICROVASCULAR EVENTS IN
THE HAMSTER CHEEK POUCH

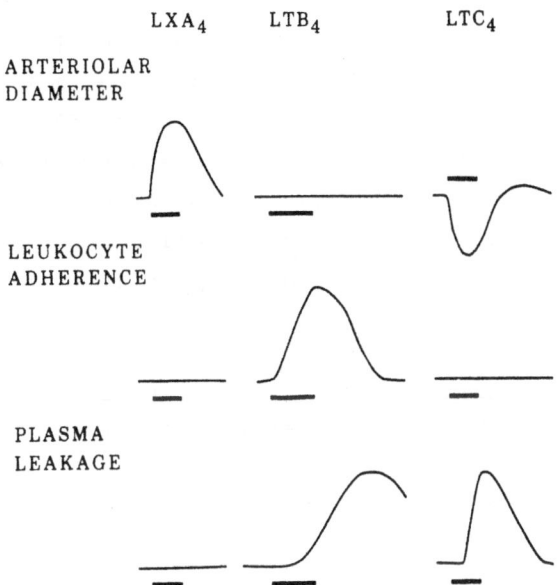

Fig. 1. Comparison of the microvascular actions of LXA$_4$ with those of LTB$_4$ and LTC$_4$. Data obtained by intravital microscopy of the hamster cheek pouch [Compiled from refs. 2,6,8.]. Thus, LXA$_4$ selectively causes arteriolar dilation, whereas LTB$_4$ has no effect on blood flow. Instead, LTB$_4$ stimulates leukocyte adherence to the venular endothelium, which results in a delayed phase of leukocyte-dependent plasma extra-vasation from venules. In contrast, LTC$_4$ has no effect on leukocyte adherence but causes arteriolar constriction and a leukocyte-independent immediate type of plasma leakage, specifically located to postcapillary venules. Vasoconstriction is however not a prerequisite for the increased microvascular permeability induced by LTC$_4$.

a bath concentration of approximately 300 nM (Fig.3.). Therefore, under the present experimental conditions, namely single non-cumulative doses given to a preparation kept at stop-flow (the influence of variations in organ bath technique is discussed in ref.7), LXA_4 was active in the same dose-range as LTB_4. However, in contrast to the twitch-like response to LTB_4 (Fig.2), the contraction induced by LXA_4 in the guinea pig lung strip was very slow in onset. In fact, there was sometimes a lag period of up to several minutes before the contraction started to develop. Moreover, the peak amplitude of the response to LXA_4(300 nM) in the lung strip was not reached until almost 20 min after administration (18.8 ± 2.3 min, mean ± SE, n=12). Taken together, the response to LXA_4 in the lung strip resembled the effects of the cysteinyl-LTs (LTC_4, LTD_4 and LTE_4) in this particular preparation. However, as indicated by the results from parallel bioassay on several preparations (Fig.2.), LXA_4 had a profile of activity which differed from that of LTC_4.

In aortic strips from rabbits or guinea pigs, LXA_4 (tested up to 1µM) did not relax preparations precontracted by norepinephrine, $PGF_{2\alpha}$ or U-44069 (TXA_2-mimetic). This was tested both in intact and endothelium-denuded preparations. Thus, in spite of its vasodilatory property in vivo (cf. above), LXA_4 was found inactive in isolated vascular strips which are sensitive to standard vasodilators such as acetylcholine or PGI_2. One explanation for this apparent discrepancy between in vivo and in vitro responses may be that LXA_4 specifically affects smooth muscle/and or endothelium in small arterioles, but not in large conducting arteries. Alternatively, species variations may contribute to the differences.

STRUCTURE-ACTIVITY RELATIONS IN SMOOTH MUSCLE

We next examined the structure activity relationship for LXA_4 in the guinea pig lung strip. The presence of an alcohol group at carbon atom 6 (C-6) was found necessary for contractile activity. Thus, neither 5S-monohydroxy

eicosatetraenoic acid (5-HETE) nor 5S,15S-dihydroxy
eicosatetraenoic acid (5,15-DHETE) had significant
contractile activity on the lung strip (Table 1.). A

PROFILES OF SPASMOGENIC
ACTIVITY IN G.P. TISSUES

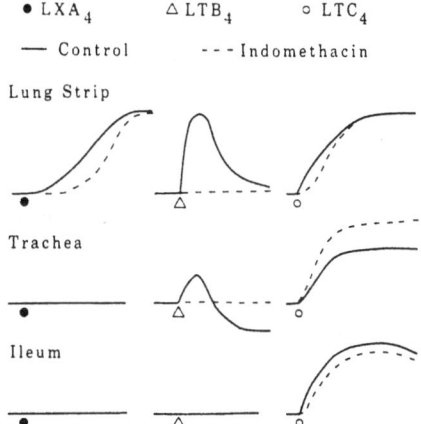

Fig. 2. Schematic drawing summarizing results obtained
in parallel assay of LXA$_4$, LTB$_4$ and LTC$_4$ on
three guinea pig prepartions, namely, lung
strips, tracheal spirals and strips of
longitudinal ileal muscle [Compiled from data in
refs. 2,7,9.]. As indicated by the broken
lines, the actions of LTB$_4$ can be explained by
release of cyclooxygenase products (in the lung
strip the constrictor thromboxane(TX) A$_2$ but in
the trachea both contractile and relaxant
prostanoids), whereas for LXA$_4$ and LTC$_4$
secondary release of prostanoids only modify the
contractions.

further indication that a hydroxyl group at C-6 is required
for the spasmogenic activity of lipoxins, was the finding
that LXB$_4$ (5S,14R,15S-trihydroxy-6,10,12-_trans_-8-_cis_-eicosa-
tetraenoic acid) but relaxed the lung strip (Table 1.).

111

Table 1. Comparisons of Contractile Activity in the Guinea Pig Lung Strip.

Compound	EC_{50} (nM)[a]	Type of response
LXA$_4$[b]	300	very slow onset, sustained,
5S-HETE	>10000	biphasic (first contraction), short-lived
5S,15S-DHETE	>10000	slow onset, but not sustained.
5,6-DHETE[c]	800	slow onset sustained
15S-HETE	6500	prompt, short-lived.
LXB$_4$[d]	–	relaxation

[a]Concentration of agonist required to achieve half maximal contractions. All responses were expressed in percent of the response to supramaximal concentrations of acetylcholine, histamine and KCl at the end of each experiment. The EC_{50}-values are approximations from non-cumulative dose-response relations established for each compound, except for 5-HETE and 5,15-DHETE, which at the highest concentration tested (10 μM) produced responses of less than 25% or maximum.
[b]5S,6R,15S-trihydroxy-7,9,13-trans-11-cis-eicosatetraenoic acid.
[c]5S,6R-dihydroxy-7,9-trans-11,14-cis-eicosatetraenoic acid.
[d]5S,14R,15S-trihydroxy-6,10,12-trans-8-cis-eicosatetraenoic acid.

On the other hand, 5,6-dihydroxy eicosatetraenoic acid (5,6-DHETE), isolated from incubations of LTA$_4$ with a recently characterized epoxide hydrolase [10], contracted the lung strip with a potency that was only slightly less than that of LXA$_4$ (Table 1.). 15S-monohydroxy eicosatetraenoic acid (15-HETE) also contracted the lung strip. However, 15-HETE was less potent than 5,6-DHETE and LXA$_4$. In addition, the response to 5,6-DHETE was of longer duration than the twitch-like contractions induced by 15-HETE (Table 1.).

112

Moreover, there were definite stereochemical requirements with respect to the orientation of the alcohol group at C-6, because 6S-LXA₄ (5S,6S,15S-trihydroxy-7,9,13-trans-11-cis-eicosatetraenoic acid) was virtually inactive on the lung strip(Fig.3.). In this context, it should be recognized that the stereochemistry of the hydroxyl groups in 5,6-DHETE appears to be 5S,6R [10]. Thus, the presence of an R-orientated alcohol group positioned at C-6 appeared crucial for spasmogenic activity. It is not known if the adjacent (5S) hydroxyl group is critical to obtain full agonist potency. However, it seems to be more than a coincidence that the same spatial arrangement of the polar substituents(5S,6R) is a prerequisite for spasmogenic activity of the cysteinyl-LTs [11,12,13]. It is therefore suggested that also in LXA₄ both the presence and the relative orientation of the two hydroxyls at C-5 and C-6 are essential for optimal recognition at a tentative receptive site.

Since LXA₄ was somewhat more potent than 5S,6R-DHETE (Table 1), it may be that the introduction of a polar group at the methyl-end of the sequence with conjugated double bonds increases agonist activity. Furthermore, in 11-trans-LXA₄ the tetraene sequence is all-trans, which means that compared with LXA₄, the hydroxyl group at C-15 is removed somewhat from the alcohols at C-5 and C-6. Therefore, the finding that 11-trans-LXA₄ was less potent than LXA₄ (Fig.3.) suggests that the presence of another polar group close to the 5S,6R-hydroxyls is advantageous but not crucial. However, 11-trans-LXA₄ retained a considerable portion of the contractile potency residing in LXA₄, indicating that the geometry of this double bond has comparatively subordinate importance for agonist activity. Considered together, the structure-activity studies indicate that the 5S,6R-hydroxyls positioned immediately adjacent to the carboxylic end of the conjugated tetraene sequence, are the primary determinants of spasmogenic activity for lipoxins.

Release of TXA$_2$

One explanation for the slow onset of the contraction evoked by LXA$_4$ in the lung strip could be that it was acting indirectly via release of myotropic factors. By measurements of TXB$_2$ release into the bath fluid, it was indeed documented that LXA$_4$ stimulated formation of TXA$_2$ in the guinea pig lung strip (Table 2.). It is well established that a number of spasmogens, including the cysteinyl-LTs, stimulate generation of TXA$_2$ from the guinea

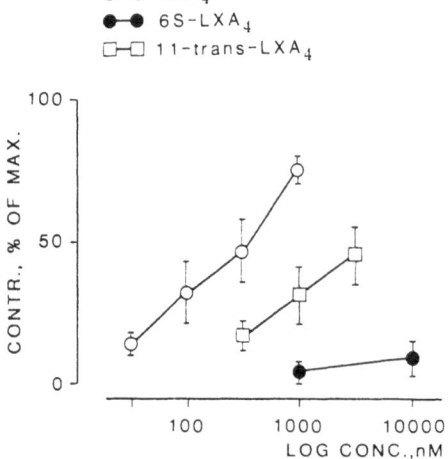

Fig. 3. Non-cumulative dose response relations for contractile activities of LXA$_4$, 11-trans-LXA$_4$ and 6S-LXA$_4$ in parenchymal strips of guinea pig lung. Mean values ± SD (% of maximal contraction) for 6-13 observations in separate tissues at each concentration.

pig lung [9,14,15]. In some instances, as exemplified by LTB$_4$, the contractile activity is due to release of TXA$_2$ (Fig.2.). On the other hand, the major portion of the

Table 2. Release of TXA_2[a] induced by LXA_4 in parenchymal strips of guinea pig lung.

	TXB_2 (pM). (Mean ± SD)
Unstimulated, 30 min (n=3)	257 ± 60
LXA_4 (3 µM) 30 min (n=10)	3194 ± 560[***]
LXA_4 (3 µM) 30 min + indomethacin (10 µM)[b] (n=4)	411 ± 40 [ns]

[a]The concentration of TXB_2 in the bath fluid was measured with a radioimmunoassay for TXB_2 (For details see ref.9). The antibody for TXB_2 did not cross-react with with LXA_4.
[b]The tissues were preincubated with indomethacin for 30 min.
[***] $p < 0.001$, as compared with unstimulated control.
[ns] Not different from control.

contraction induced by LTC_4 is TXA_2-independent (Fig.2.), although the relative importance of TXA_2-release varies with the experimental conditions [7].

In the case of LXA_4, 30 min pretreatment with indomethacin (10 µM) blocked release of TXA_2 (Table 2.) but did not reduce the peak amplitude of the response to LXA_4 (Fig.4.). However, after indomethacin, there was a significant delay of the onset of the contraction, and at time points up to 20 min after challenge, indomethacin depressed the time-tension curve for LXA_4. Therefore, to some extent, TXA_2 contributed to the early part of the contraction induced by LTC_4. Nevertheless, it is evident that under the present experimental conditions, release of the TXA_2 cannot explain the spasmogenic action of LXA_4 in the guinea pig lung strip. Likewise, there are no indications that acetylcholine, norepinephrine or histamine contribute to the contractions induced by LXA_4 in the lung strip, because the response to LXA₄ was found resistant to appropriate receptor-antagonist for these autacoids [2].

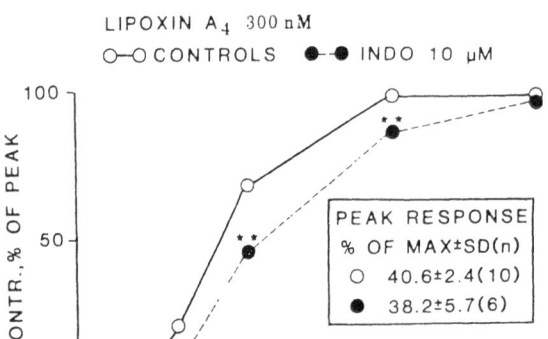

TIME-COURSE OF CONTRACTION

LIPOXIN A₄ 300 nM
○—○ CONTROLS   ●—● INDO 10 µM

Fig. 4. Effect of 30 min pretreatment with indomethacin (INDO) on the contraction induced by LXA₄ in the guinea pig lung strip. As indicated, the peak amplitude of the contraction was not reduced by indomethacin, but the time required to reach the peak value was prolonged.

Pharmacological antagonism of LXA₄

In view of the similarities between LTC₄ and LXA₄ with respect to the time-course of the contraction, the structure-activity relationship, as well as the ability to stimulate generation of TXA₂ from the guinea pig lung, the influence of FPL 55712 on responses to LXA₄ was examined (Fig.5.). This effector antagonist for cysteinyl-LTs [16] unequivocally blocked the spasmogenic activity of LXA₄. The antagonism could not be accounted for by unspecific

116

depression of tissue reactivity, because the maximal
contractility of the preparations were not altered, nor did
FPL 55712 at the concentration used affect responses induced
by histamine or $PGF_{2\alpha}$ in the lung strip (Fig.5.).

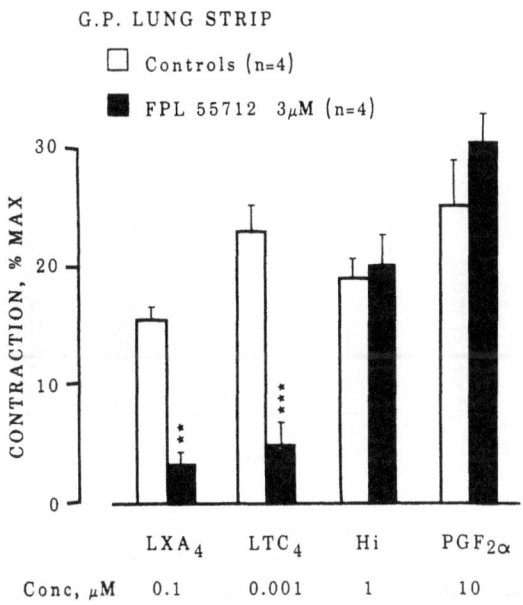

Fig. 5. The antagonist for cysteinyl-LTs, FPL 55712
        blocked responses in the lung strip evoked by
        LXA_4 and LTC_4 without affecting contractions
        elicited with histamine (Hi) or $PGF_{2\alpha}$.

The findings with FPL 55712 raised two possibilities.
Lipoxin A_4 could either be acting at the same site at the
effector cell as the cysteinyl-LTs, or, alternatively, LXA_4
could provoke the liberation of leukotrienes from the lung
parenchyma. The experimental data (Fig.6.) supported the
first possibility, namely that FPL 55712 antagonized LXA_4
directly at the effector cell level. Thus, two structurally
unrelated lipoxygenase inhibitors failed to affect the

contraction induced by LXA₄ in the lung strip, whereas two other effector-antagonists for cysteinyl-LTs shared the ability of FPL 55712 to block contractions induced by either LTC₄ or LXA₄ (Fig.6). In fact, the effector antagonists seemed more effective against LXA₄ than LTC₄.

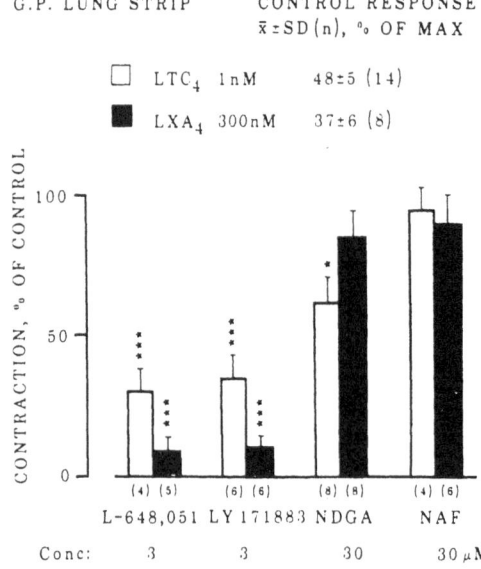

Fig. 6. The LT-antagonists L-648,051 [17] and LY-171883 [18] blocked responses to equiactive concentrations of LXA₄ and LTC₄ in the guinea pig lung strip. In contrast, Nafazatrom (NAF) [19] and Nordihydroguaiaretic acid (NDGA) [20], two lipoxygenase inhibitors, had no or comparatively minor effects on contractions evoked by LXA₄ and LTC₄.

If LXA₄ universally affected smooth muscle at a site with closely similar characteristics as the tentative receptors for cysteinyl-LTs, it is hard to explain why LXA₄ was inactive on the guinea pig ileum and trachea (Fig.2.). Since the ileum is less sensitive to cysteinyl-LTs than the lung strip, one possibility was that the initial study performed with biologically derived LXA₄ [2], had employed subthreshold doses. However, reinvestigations with synthetic material showed that LXA₄ failed to contract the

118

guinea pig ileum although tested in concentrations up to 30 μM. This concentration is 10000 times above the threshold for LTC$_4$ or LTD$_4$ in the ileum [7]. As a comparison, in the lung strip, the potency ratio between LTC$_4$ and LXA$_4$ is approximately 1:100 [2,9]. Therefore, although both the lung strip and the ileum are sensitive to LTC$_4$, the receptors in the two tissues obviously differ with respect to sensitivity for LXA$_4$. The present observations with LXA$_4$ may thus relate to other indications that there are multiple and tissue-specific receptors for cysteinyl-LTs [7,11,21].

Interactions between LXA$_4$ and LTC$_4$

Guinea pig lung strip. To further explore the indications that LXA$_4$ affected sensitive smooth muscle preparations by an action located at the same or a closely similar site as the cysteinyl-LTs, it was investigated whether or not synergism existed between LTC$_4$ and LXA$_4$ in the lung strip.

This was however not the case. If a cumulative dose-response relation for LTC$_4$ was established in the presence of a low dose of LXA$_4$, there was but an additive effect (Fig.7A.). A low dose of LTD$_4$ also shifted the dose-response relation for LTC$_4$ in an identical additive manner (not shown). On the other hand, histamine, which clearly acts at another receptor than the cysteinyl-LTs, failed to alter the dose-response relation for LTC$_4$ (not shown).

In addition, if a preparation was exposed to a high dose of LXA$_4$ for 30 min, the reactivity to LTC$_4$ was markedly depressed (Fig.7B.). Interestingly enough, this type of desensitization may be induced by LTC$_4$ itself [9]. Furthermore, we have recently observed another similarity between the cysteinyl-LTs and LXA$_4$. Namely, a cumulative dose-response relation for LXA$_4$ is shifted considerably to the right as compared with the relationship established with non-cumulative observations accumulated from a large sample of tissues each being exposed to only one dose of LXA$_4$.

The same phenomena is known for the cysteinyl-LTs . In view of the experience gathered, with leukotrienes over the past ten years, one may thus envisage that experimentation with lipoxins in smooth muscle preparations can be

G.P. LUNG STRIP

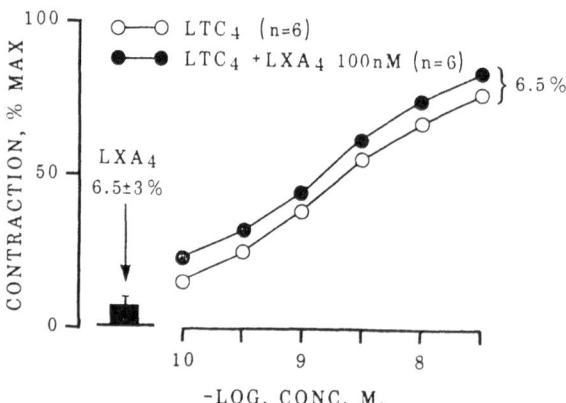

Fig. 7A. Interactions between LXA$_4$ and LTC$_4$ in the guinea pig lung parenchyma. Cumulative dose-response relations for the contractile activity of LTC$_4$ were established in preparations exposed to LXA$_4$ and parallel (untreated) controls. In this experiment, a low dose of LXA$_4$ was added 15 min before LTC$_4$ and remained in the bath during challenge with LTC$_4$. The filled bar to the left of the dose-response curves indicate the response to LXA$_4$ (values in figure are mean contraction ± SE, % of the maximum). Stars indicate level of significance (* = $p < 0.05$, etc).

complicated if the presence of desensitizations and interactions are overlooked. Taken together, these observations indicate that in the lung strip preparation, LXA$_4$ shared the same receptor and/or transduction mechanism as the cysteinyl-LTs.

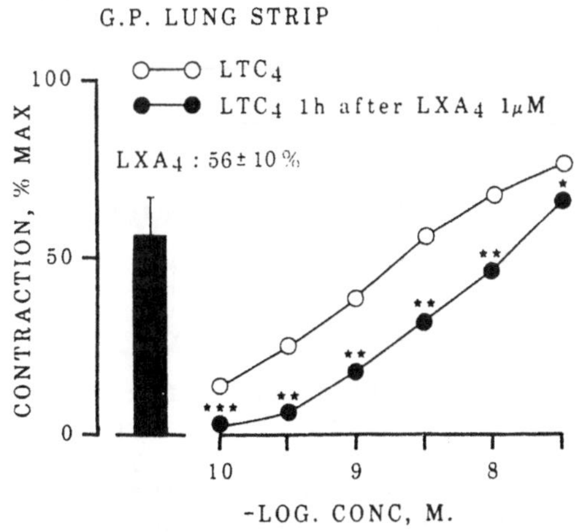

Fig. 7B. Interactions between LXA$_4$ and LTC$_4$ in the guinea pig lung parenchyma. In this experiment, a high dose of LXA$_4$ was given 90 min before the cumulative challenge with LTC$_4$. Lipoxin A$_4$ remained in contact with the preparation for 30 min. One hour after LXA$_4$ had been washed out, the preparation had normal reactivity to histamine (not shown) but the dose-response relation for LTC$_4$ was depressed. (For details see legend to Fig. 7A.).

Guinea pig ileum. Although LXA₄ by itself lacked
contractile activity (see above), it was found that LXA₄
dose-dependently depressed the response to LTC₄ in the ileum
(Fig.8.). At the concentration of 1 µM, LXA₄ caused a
35 ± 8% (mean ± SE, n=10, p<0,05) inhibition of the response
to LTC₄ (1 nM). In addition, when LXA₄ was given to a
preparation precontracted by LTC₄, the contraction was
readily reversed, similar to the effect of FPL 55712 in this
preparation. In this respect, LXA₄ was only one half log

G.P. ILEUM LONGITUDINAL MUSCLE

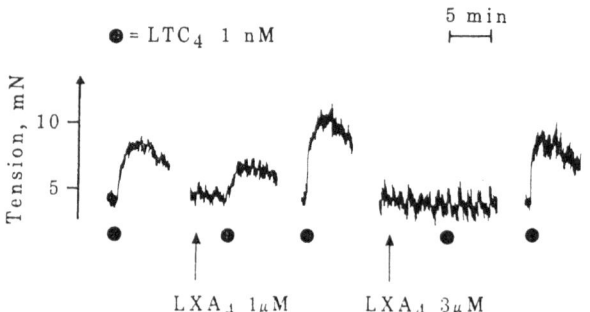

Fig. 8. Representative tracings of contractions in the
longitudinal muscle of the guinea pig ileum.
Leukotriene C₄ added every 30 min (dots),
pretreatment with LXA₄ indicated by arrows. The
solvent (ethanol, less than 0.1% final bath
concentration) neither affected basal tone nor
the response to histamine or LTC₄.

order less active than FPL 55712. The cause of this
difference between the influence of LXA₄ on leukotriene-
responses in the ileum and the lung strip remains to be
explained, but the observations lend further support to the
view that receptors for leukotrienes differ considerably
between these two commonly employed preparations.

Guinea pig trachea. Five min pretreatment with LXA$_4$ (1 μM) depressed the contraction induced by LTC$_4$ (10 nM) by approximately 25 ± 12% (Means ± SD, n=9, P<0.05), as compared with parallel controls being challenged in the absence of LXA$_4$. The nature of this interaction is subject to further investigations.

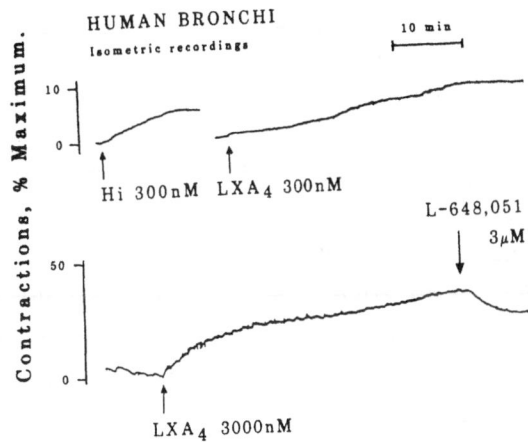

Fig. 9. Tracings of contractions induced by LXA$_4$ and histamine (Hi) in isolated human bronchi. Note the time-scale. L-648,051 is an effector antagonist for cysteinyl-LTs (see text and ref. 19.).

Action of LXA$_4$ on Human Bronchi

LXA$_4$ displayed contractile activity also on human bronchi (Fig.9.). The time-course of the response was however even slower than in the guinea pig lung strip, and sometimes more than an hour was required before the peak of the contraction was reached. Moreover, the effector antagonist for cysteinyl-LTs, L-648,051, was able to reverse the contraction induced by LXA$_4$. Thus, the experiments suggested that LXA$_4$ in human airways acted by mechanisms similar to those observed in the guinea pig lung strip.

CONCLUSIONS

As outlined in the present chapter, LXA$_4$ displayed biological activity in the pulmonary and microvascular systems. Thus, it caused contractions of human bronchi and parenchymal strips of guinea pig lung, dilation of the terminal vascular bed in the hamster cheek pouch, and release of TXA$_2$ from the guinea pig lung. In addition, in selected smooth muscles, LXA$_4$ had the capacity to modulate the response to LTC$_4$ at the effector cell level.

It is certainly premature to make functional interpretations of the present observations, but it is of interest that lipoxygenation of arachidonic acid at C-15 and C-5 is extensive in the human lung [22,23]. In addition, there are indications that 15-HETE is formed in the vascular endothelium [24]. Moreover, so far biosynthesis of lipoxins has been described in leukocytes, and most tissues would clearly seem to have the potential to be exposed to lipoxins from this source.

The action of LXA$_4$ on the microvasculature of the hamster cheek pouch differed from the activities known for the leukotrienes. Although LXA$_4$ apparently was inactive on vascular strips isolated from large arteries, it caused a prominent dilation of small arterioles in the hamster cheek pouch, suggesting that it had a specific inclination to act on the microvasculature. Accordingly, in the rat glomerular microcirculation, it has been observed that LXA$_4$ selectively dilates preglomerular arterioles [25].

The finding that LXA$_4$ promoted release of TXA$_2$ from the guinea pig lung, could not explain its contractile activity in the lung strip preparation. However, it is reasonable to assume that, the observation with LXA$_4$ in the guinea pig lung, reflects a general capacity for this compound to stimulate formation of cyclooxygenase products. Therefore, it is possible that, in other systems, LXA$_4$ may mediate important responses via release of cyclooxygenase products.

Furthermore, there were highly specific stereochemical requirements for the contractile activity of lipoxins in

124

sensitive smooth muscles. For example, in the guinea pig lung strip, LXA₄ was the most potent agonist whereas 6S-LXA₄ was virtually inactive. With respect to biological function, it is of interest that formation of LXA₄ predominated over 6S-LXA₄ in activated human leukocytes [26]. From the structure-activity studies it was concluded that the 5S,6R-orientation of the two hydroxyls positioned immediately adjacent to the carboxylic end of the conjugated tetraene was essential for the contractile activity of LXA₄. Interestingly enough, this trait is strikingly similar to the requirements with respect to polar substituents of cysteinyl-LTs [11,12,13].

Additional experiments utilizing different pharmacological tools indeed suggested that LXA₄ directly activated sensitive smooth muscles at a site with similar characteristics as the tentative receptor(s) for cysteinyl-LTs. Thus, three different antagonists for cysteinyl-LTs (FPL 55712, LY-171883 and L-648,051) blocked the contraction response to LXA₄ whereas inhibitors of lipoxygenases left the contraction unaltered.

The hypothesis that LXA₄ could interact with cysteinyl-LTs was tested with LTC₄ and found correct. Namely, in the guinea pig lung parenchyma, the presence of cross-tachyphylaxis and direct additive effects between LXA₄ and LTC₄ indicated specific interactions at the receptor-level. In the guinea pig ileum and trachea, however, the interaction involved inhibition of responses to LTC₄ by LXA₄. The interactions between LXA₄ and LTC₄ suggest that both up- and down-regulation of effector cell responsiveness may be achieved within the lipoxygenase system.

In addition, the variations in sensitivity to LXA₄ between smooth muscles of different origin is in line with other indications that there are multiple forms of receptors for cysteinyl-LTs. It is evident that LXA₄ is an agonist at the receptor for cysteinyl-LTs in the guinea pig lung strip but fails to activate the type of receptor present in the ileum. On a speculative note, it may be suggested that LXA₄ is a partial agonist at the leukotriene-receptor in the ileum. The pharmacological analysis required to test this possibility may provide interesting information about the mode of action for leukotrienes, lipoxins, as well as effector-antagonists for leukotrienes.

Although LXA₄ interacted with leukotrienes in smooth muscle, it cannot be concluded that lipoxins generally have this mode of action. In addition to the spasmogenic and microvascular properties of LXA₄ discussed in this chapter, LXA₄ has been found to affect cellular functions such as secretion from leukocytes [27], cytotoxicity induced by natural killer cells [28,29] and activation of protein phosphorylation by protein kinase C [30]. The information that accumulates in fact suggests considerable variations in structure-activity requirements for the different effects of lipoxins. For example, in contrast to the lung strip, LXA₄ and LXB₄ have similar activities in the cytotoxicity assay. With respect to activation of protein kinase C, LTC₄ was less potent than LXA₄. Furthermore, the responses to 15-HPETE and LXA₄ were similar in the hamster cheek pouch but very different in the guinea pig lung strip.

Finally, in addition to differences in structure-activity requirements for the various biological activities described for lipoxins, it should also be recognized that within each assay system, the mode of action or the profile of activities displayed by lipoxins were often distinct from those of other autacoids. For example, in the spasmogenic assays and on the microcirculation, the actions of LXA₄ were distinguished from those exerted by leukotrienes, thromboxanes and prostaglandins. Thus, it seems justified to conclude that on the basis of pharmacological evidence, the lipoxins have the potential to subserve specific regulatory functions distinct from those of other eicosanoids. When more is known about conditions for lipoxin formation in vivo, it will be possible to elucidate whether or not lipoxins function as mediators of inflammation, modulators of immune competence, intracellular messengers, or have other roles.

ACKNOWLEDGMENTS

We are grateful to Dr. J. Rokach (Merck-Frosst Inc, Canada) for a generous supply with synthetic leukotrienes and L-648,051, Dr. J. Pike (Upjohn Co, USA) for U-44069, Dr. D.K. Rainey, Fisons plc (U.K.) for FPL 55712, Bayer AG (West Germany) for Nafazatrom, Eli Lilly & Co (USA) for LY-171883, and Dr.R. Zipkin (Biomol, USA) for an aliquot of synthetic lipoxins. This work was supported by the Swedish MRC (03P-6949, 14X-4342 and K87-03R-8110 to S.-E.D.; 03X-217 to B.S.), the Swedish Association Against Chest and Heart Diseases, the National Institute of Environmental Medicine, Knut and Alice Wallenberg Foundation, Harald and Greta Jeansson Foundation and Karolinska institutet. C.N.S. is a recipient of the J.V. Satterfield Arthritis Investigator Award from the American Arthritis Foundation.

REFERENCES

1. C.N. Serhan, P. Fahlstadius, S.-E. Dahlén, M. Hamberg, and B. Samuelsson, Biosynthesis and Biological Activities of Lipoxins, in: "Advances in Prostaglandin, Thromboxane and Leukotriene Research," vol.15, O. Hayaishi and S.Yamamoto, eds, Raven Press, New York (1985).

2. S.-E. Dahlén, J. Raud, C.N. Serhan, J. Björk, and B. Samuelsson, Biological activities of Lipoxin A include lung strip contraction and dilation of arterioles in vivo, Acta Physiol Scand 130:643 (1987).

3. K.C. Nicolaou, C.A. Veale, S.E. Webber, and H. Katerinopoulos, Stereocontrolled total synthesis of lipoxins A, J Amer Chem Soc, 107:7515 (1985).

4. J. Björk, G. Smedegård, E. Svensjö, and K.-E. Arfors, The use of the hamster cheek pouch for intravital studies of microvascular events, Progr Appl Microcirc, 6:41 (1984).

5. J. Raud, S.-E.Dahlén, A.Sydbom, L.Lindbom and P.Hedqvist. Enhancement of acute allergic inflammation by indomethacin is reversed by Prostaglandin E_2: Apparent correlation with in vivo modulation of mediator release. Proc Natl Acad Sci USA,(In press).

6. S.-E. Dahlén, J. Björk, P. Hedqvist, K.-E. Arfors, S. Hammarström, J.A. Lindgren, and B. Samuelsson, Leukotrienes promote plasma leakage and leukocyte adhesion in postcapillary venules: in vivo effects with relevance to the acute inflammatory response, Proc Natl Acad Sci USA 78:3887 (1981).

7. S.-E. Dahlén, T. Björck, and P. Hedqvist, Bioassay of Leukotrienes, in: "Advances in Prostaglandin, Thromboxane and Leukotriene Research," vol.17, B. Samuelsson, R. Paoletti and P. Ramwell, eds., Raven Press, New York (1987).

8. J. Björk, P. Hedqvist, and K.-E. Arfors, Increase of vascular permeability induced by leukotriene B_4 and the role of polymorphonuclear leukocytes, Inflammation, 6:189 (1982).

9. S.-E. Dahlén, P. Hedqvist, P. Westlund, E. Granström, S. Hammarström, J.A. Lindgren, and O.Rådmark, Mechanisms for leukotriene-induced contractions of guinea pig airways: Leukotriene C_4 has a potent direct action whereas leukotriene B_4 acts indirectly, Acta Physiol Scand, 118:393 (1983).

10. J. Haeggström, J. Meijer, and O, Rådmark, Leukotriene A_4, enzymatic conversion into 5,6-dihydroxy-7,9,11,14-eicosatetraenoic acid by mouse liver cytosolic epoxide hydrolase, J Biol Chem. 261:6332 (1986).

11. R.A. Lewis, J.M. Drazen, K.F. Austen, M. Toda, F. Brion, A. Marfat, and E.J. Corey, Contractile activities of structural analogs of leukotrienes C_4 and D_4: Role of the polar substituents, Proc Natl Acad Sci USA, 78: 4579 (1981).

12. S.R. Baker, J.R. Booth, W.B. Jamieson, D.J. Osborne, and W.J.F. Sweatman, The comparative in vitro pharmacology of leukotriene D_4 and its isomeres, Biochem Biophys Res Commun, 103:1258 (1981).

13. S.-E. Dahlén, P.Hedqvist, and S.Hammarström, Contractile activities of several cysteine-containing leukotrienes in the guinea-pig lung strip, Eur J Pharmacol, 86:207 (1983).

14. G.C. Folco, G. Hansson, and E.Granström, Leukotriene C_4 stimulates TXA_2 formation in isolated sensitized guinea pig lungs, Biochem Pharmacol, 30:2491 (1981).

15. P.J. Piper, and M.N. Samhoun, Stimulation of arachidonic acid metabolism and generation of thromboxane A_2 by leukotrienes B_4, C_4 and D_4 in guinea-pig lung in vitro, Br J Pharmacol, 77:267 (1982).

16. J. Augstein, J.B. Farmer, T.B. Lee, P. Sheard, and M.L. Tattersall, Selective inhibitor of slow reacting substance of anaphylaxis, Nature New Biol, 245:215 (1973).

17. T.R. Jones, Y. Guindon, R. Young, E. Champion, L. Charette, D. Denis, D. Ethier, R. Hamel, A.W. Ford-Hutchinson, R. Fortin, G. Letts, P. Masson, C. McFarlane, H. Piechuta, J. Rokach, C. Yoakim, R.N. De Haven, A. Maycock, and S.S. Pong, L-648,051, Sodium 4-[3-(4-acetyl-3-hydroxy-2-propylphenoxy)-propyl-sulfonyl]-r-oxo-benzenebutanoate: a leukotriene D_4 receptor antagonist, <u>Can J Physiol Pharmacol</u> 64:1535 (1986).

18. J.H. Fleisch, L.E. Rinkema, K.D. Haisch, D. Swanson-Bean, T. Goodson, P.P.K. Ho, and W.S. Marshal, LY171883, 1-<2-hydroxy-3-Propyl-4-<4-(1H-Tetrazol-5-yl)Butoxy>Phenyl>Ethanone, an Orally Active Leukotriene D_4 Antagonist, <u>J Pharm Exp Ther</u>, 233:148 (1986).

19. W.-D. Busse, M. Mardin, R. Grützman, J.R. Dunn, M. Theodorou, B.F. Sloane, and K.V. Honn, Nafazatrom (BAY G 6575), an inhibitor of cellular lipoxygenase activity, <u>Fed Proc</u>. 41:1717 (1982).

20. M. Hamberg, On the formation of thromboxane B_2 and 12L-hydroxy-5,8,10,14-eicosatetraenoic acid (12ho-20:4) in tissues from the guinea pig, <u>Biochim Biophys Acta</u>, 431:651 (1976).

21. J.H. Fleisch, L.E. Rinkema, and W.S. Marshall, Pharmacologic Receptors for Leukotrienes, <u>Biochem Pharmacol</u>, 33:3919 (1984).

22. M. Hamberg, P.Hedqvist, and K.Rådegran, Identification of 15-hydroxy-5,8,11,13-eicosatetraenoic acid (15-HETE) as a major metabolite of arachidonic acid in human lung, <u>Acta Physiol Scand</u>, 110:219 (1980).

23. S.-E. Dahlén, G. Hansson, P. Hedqvist, T. Björck, E. Granström, and B. Dahlén, Allergen challenge of lung tissue from asthmatics elicits bronchial contraction that correlates with the release of leukotrienes C_4, D_4 and E_4, <u>Proc Natl Acad Sci USA</u>, 80:1712 (1983).

24. R.R. Gorman, T.D. Oglesby, G.L. Bundy, and N.L. Hopkins, Evidence for 15-HETE synthesis by human umbilical vein endothelial cells, <u>Circulation</u>, 72:708 (1985).

25. K.F. Badr, C.N. Serhan, K.C. Nicolaou, and B.Samuelsson, The action of lipoxin-A on glomerular microcirculatory dynamics in the rat, <u>Biochem Biophys Res Commun.</u>, 145:408 (1987).

26. C.N. Serhan, K.C. Nicolaou. S.E. Webber, C.A. Veale,

S.-E. Dahlén, T.J. Puustinen, and B. Samuelsson, Lipoxin A: Stereochemistry and biosynthesis, <u>J Biol Chem</u>. 261:16340 (1986).

27. C.N. Serhan, M. Hamberg, and B. Samuelsson, Lipoxins: Novel series of biologically active compounds formed from arachidonic acid in human leukocytes, <u>Proc Natl Acad Sci USA</u>, 81:5335 (1984).

28. U. Ramstedt, J. Ng, H. Wigzell, C.N. Serhan, and B. Samuelsson, Action of Novel Eicosanoids Lipoxin A and B On human natural Killer Cell Cytotoxicity: Effects on intracellular cAMP and Target Cell Binding, <u>J.Immunol</u>. 135:3434 (1985).

29. U. Ramstedt, C.N. Serhan, K.C. Nicolaou, S.E. Webber, H. Wigzell, and B. Samuelsson, Lipoxin A-induced inhibition of natural killer cells: Studies on stereospecificity and mode of action, <u>J Immunol</u>. 138:266 (1987).

30. A. Hansson, C.N. Serhan, J. Haeggström, M. Ingelman-Sundberg, B. Samuelsson, and J. Morris, Activation of protein kinase C by lipoxin A and other eicosanoids. Intracellular action of oxygenation products of arachidonic acid, <u>Biochem Biophys Res Commun</u>. 134:1215 (1986).

10

THE GLOMERULAR PHYSIOLOGY OF LIPOXIN-A

Kamal F. Badr

Renal Division, Department of Medicine
Vanderbilt University
Nashville, TN 37232

INTRODUCTION

A number of renal diseases, such as acute and chronic glomerulo-nephritis, acute interstitial nephritis, allograft rejection, and others are characterized by the presence inflammatory cellular infiltrates consisting, at various stages, of neutrophils, macrophages, monocytes, eosinophils, and lymphocytes. In addition, these disease processes are often accompanied by impairment of glomerular perfusion, filtration, and permselectivity functions. Despite much controversy as to the mediator systems involved in the initiation and perpetuation of these glomerular functional abnormalities, consensus has emerged as to the central role of leukocyte-derived biologically active cytokines[1]. In this regard, lipid-derived mediators, including platelet activating factor and the cyclooxygenase and lipoxygenase (LO) products of arachi-donate metabolism have received particular attention. An understanding of the role played by these locally released compounds in the patho-physiology of glomerular injury can only result from a close integration of the best available bioanalytic and physiologic technologies, and the application of both approaches to experimental disease models, in conjunction with selective antagonism of specific mediators through receptor antagonists and enzyme inhibitors. Using glomerular micro-puncture techniques, we have identified the responses of the glomerular microcirculation to peptidyl leukotrienes[2,3] and, utilizing selective antagonists, proposed a role for these compounds in mediating the reductions in renal plasma flow and glomerular filtration, as well as the induction of proteinuria, in endotoxin- and anti-glomerular basement membrane antibody-induced glomerular injury[4,5]. In more recent studies[6], we have examined the effects of exogenously administered lipoxin A (LX-A) on the glomerular microcirculation in the rat. Our results indicate that, in sharp contrast to sulfidopeptide LTs, the predominant action of LX-A is to increase renal perfusion. This response appears to be evoked by selective reduction in preglomerular resistance with a consequent aug-mentation of single nephron and whole kidney glomerular filtration rate.

METHODS

Biologically derived LX-A was isolated and purified from activated leukocytes as described[7,8]. The Me_3Si derivative of this material eluted as a major component on GC with a C-value of 24.1. The prominent ions

were at m/e 203 (base peak), 171, 173, 289, and 379. Ions at lower
intensity were at 402, 482, 492, and 582 (M). These ions and C-values
are identical to those reported for LX-A[7,8]. Synthetic lipoxin A was
prepared as in [9]. The synthetic and biologically derived materials were
matched by published criteria[7]. In the present study, the biologically
derived and synthetic materials proved to display identical biological
actions. LX-A free acid or methyl ester aliquotes were stored in ethanol
under argon at -70°C. On the day of the experiment, ethanol was evaporated
and the lipoxin resuspended in 0.9% NaCl at the appropriate dilution.

All experiments were performed on anesthetized adult male Munich-
Wistar rats weighing 175-230 gms which were prepared for micropuncture
according to protocols described previously[10]. In brief, following
Inactin anathesia (100 mg/Kg, i.p), the left femoral artery was catheter-
ized and used to monitor mean systemic arterial pressure (AP) and for
sampling of blood.

Following a tracheostomy, catheters were inserted into both jugular
veins for infusion of plasma and inulin (7.5% solution in 0.9% NaCl at
1.2 ml/hr). The left kidney was exposed by a left subcostal incision,
separated from the surrounding fat, and suspended on a Lucite holder.
The kidney surface was illuminated with a fiberoptic light source and
bathed with isotonic NaCl. A 30-gauge needle was placed in the abdominal
aorta at the take-off of the left renal artery through which a maintenance
infusion of 0.9% NaCl at the rate of 0.025 ml/min was initiated. Also,
an electromagnetic flow probe was placed round the left renal artery and
connected to a flow meter (Carolina Medical Electronics) which allowed
for continuous monitoring of renal blood flow rate. Homologous rat plasma
was administered intravenously according to a protocol shown previously
to adequately replace surgically-induced plasma losses, thus maintaining
euvolemia[10].

In all experiments, micropuncture measurements were started 60 min
after the onset of plasma infusion and carried out as follows: Exactly
timed (1-2 min) samples of fluid were collected from surface proximal
convolutions of each of two to three nephrons for determination of flow
rate and inulin concentration and calculation of tubule fluid-to-plasma
inulin concentration ratio and single nephron glomerular filtration rate
(SNGFR). Coincident with these tubule fluid collections, two or three
samples of femoral arterial blood were obtained in each period for
determination of systemic arterial hematocrit (Hct) and plasma concentration of total protein and inulin. Also, at least three samples of
blood were obtained from surface efferent arterioles for determination
of efferent arteriolar protein concentration. In addition, two or three
samples of urine from the experimental kidney were collected for the
determination of flow rate and inulin concentration, and for the
calculation of whole kidney GFR. For these urine collections, indwelling
polyethylene ureteral catheters were used (PE 10).

Time-averaged hydraulic pressures were measured in surface glomerular
capillaries (P_{GC}), proximal tubules (P_T), and surface efferent arterioles
(P_E) using a continuous recording, servo-null micropipette transducer
system (Model 5, Instrumentation for Physiology and Medicine, San Diego,
CA). Micropipettes with outer tip diameters of 2-4 µm and containing
2.0 M NaCl were used. Experiments were performed on two groups of rats:

Group I (n=5): In this group, whole kidney and micropuncture
measurements were performed during an initial baseline period and then
repeated during a second period in which the LX-A vehicle infusion (0.9%
NaCl) into the left renal artery was maintained. These animals served
as time controls.

Group II (n=10): In this group, whole kidney and micropuncture measurements were performed during an initial baseline period and then repeated during intra-renal arterial administration of LX-A in a dose of 750 ng/kg/min, which was maintained for approximately 30 min. at a rate equal to 0.025 ml/min.

ANALYTICAL

Colloid osmotic pressures of plasma entering and leaving glomerular capillaries were estimated from values for protein concentrations (C) in femoral arterial (C_A) and surface efferent arteriolar (C_E) blood plasmas. Colloid osmotic pressure (π) was calculated according to the equation derived by Deen et al. [11]. Values for C_A, and thus π_A, for femoral arterial plasma are taken as representative for values for C and π for the afferent end of the glomerular capillary network. These estimates of pre- and postglomerular protein concentration permit calculation of single nephron filtration fraction (SNFF), and glomerular capillary ultrafiltration coefficient (K_f), as well as resistance of single afferent (R_A) and efferent (R_E) arterioles, total arteriolar resistance (R_{TA}), and initial glomerular capillary plasma flow (Q_A), using equations described in detail elsewhere[11].

The volume of fluid collected from individual proximal tubules was estimated from the length of the fluid column in a constant bore capillary tube of known internal diameter. The concentration of inulin in tubule fluid was measured by the microfluorescence method of Vurek and Pegram[12]. Inulin concentrations in plasma and urine were determined by the macroanthrone method of Fuhr et al.[13]. Protein concentration in efferent arteriolar and femoral arterial blood plasmas were determined using a fluorometric method developed by Viets et al.[14].

Statistical: Paired t-test was performed to compare, within each group, the changes in various whole kidney and microcirculatory indices which occurred from the first to the second study period. Differences were considered significant at a p value $\leqq 0.05$. All values are reported as mean ± SEM.

RESULTS

In Group I animals, no significant changes were noted between the first and second period in any of the systemic or renal parameters monitored. In addition, baseline values for these parameters in Group I rats were not statistically different from those of Group II animals described below.

In Group II rats, administration of LX-A was not associated with significant changes in AP or Hct. Despite constancy of these systemic parameters, GFR increased from 1.00±0.7 to 1.33±0.12 ml/min (p<0.05) as did renal plasma flow (RPF) which increased from 3.09±0.21 to 3.91±0.22 ml/min (p<0.05). Micropuncture measurements revealed similarly significant increases in SNGFR and Q_A from 38.4±1.7 to 45.5±3.0 and from 95±6 to 127±9 nl/min, respectively. The increase in Q_A was due to a selective fall in pre-glomerular (afferent) arteriolar resistance (R_A) which fell from 2.55±0.16 to 1.75±0.20 10^{10}dyn s cm^{-5} (p < 0.05), while post-glomerular (efferent) arteriolar resistance (R_E) was unchanged from 1.70±0.22 to 1.50±0.08 10^{10}dyn s cm^{-5}. The constancy of R_E, coupled with the increased Q_A, resulted in a significant rise in intraglomerular capillary hydraulic pressure (P_{GC}) from 46±2 to 53±3 mmHg (p< 0.05) which, in conjunction with a small, but significant fall in P_T (13±1 to 10±1 mmHg, p<0.05) led to an increase in ΔP from 33±1 to 43±3 mmHg (p<0.05). The increases in glomerular perfusion (Q_A) and the net ultrafiltration pressure (ΔP) were jointly responsible for the observed LX-A-induced increase in single nephron and whole kidney GFR. The latter effect, however, was partially offset by a concomitant significant reduction in the glomerular capillary ultrafiltratin coefficient (K_f) which fell from 0.060 ± 0.013 to 0.033 ± 0.005 nl/(s mmHg) ($p < 0.05$). Figure 1 summarizes the effects of LX-A on the glomerular microcirculation.

DISCUSSION

In these experiments LX-A was associated with dramatic and selective relaxation of pre-glomerular resistance vessels in the Munich-Wistar rat. The resultant glomerular hyperperfusion and hypertension led to a significant augmentation of the glomerular filtration rate. These vasorelaxant effects of this eicosanoid are in sharp contrast to those reported previously for the sulfidopeptide leukotrienes[2,3,15]. The latter, possess potent vasoconstrictor effects on arteriolar and mesangial smooth muscle[3,16,17] and their exogenous administration results in marked renal vasoconstriction and reduction in GFR[2,3,15]. A potentially central role for these compounds in mediating the reductions in glomerular perfusion and filtration during the course of experimental glomerulonephritis has been proposed recently[6]. The present experiments raise the interesting possibility that the net functional response to the activation of the LO pathways during inflammatory injury may depend, in part, on the relative predominance of the biological activities of their principal vasoactive end-products: leukotrienes and lipoxins. Recent studies indicate that 15-hydroxyeicosatetraenoic acid (15-HETE) can be transformed to lipoxins by activated human leukocytes[7] suggesting that cell-cell interactions may play a role in the generation of lipoxins and related compounds. In this context, it has been reported that the kidney is rich in 15-LO activity and that the cell types of this organ can interact in the trans-cellular formation of LTs and other novel eicosanoids[18]. In addition, the interest in a role for lipoxins and related compounds in inflammatory injury is further highlighted by the observations of Soberman et al.[19] that the leukocyte 15-LO is preferentially activated under specific conditions, and by the recent demonstration by Vanderhoek et al[20] that endogenously produced mono-hydroxyeicosatetraenoic acids are potent stimulants of the 15-LO pathway in the human leukocyte.

The response to LX-A of the glomerular microcirculation observed in the present study could well involve the release of other mediators by LX-A. Although this issue is not addressed in the present report, its relevance is highlighted by the effect of LX-A on K_f. The reduction in this parameter, usually interpreted as representing contraction of mesangial cell smooth muscle elements, contrasts with the relaxant effect of LX-A on the afferent arteriole, and raises the possibility that LX-A may stimulate the release of a K_f-lowering mediator.

In summary, these experiments demonstrate that LX-A induces glomerular hyperperfusion, hypertension, and hyperfiltration. These effects are in sharp contrast to those of the LTs in this system, and may represent the first demonstration of counterregulatory (costrictor/ dilator) vascular interactions within the two major LO pathways (5 and 15-LO).

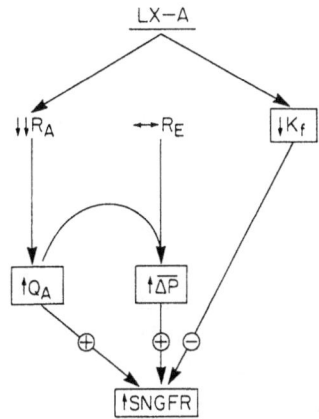

Figure 1. The actions of LX-A on the glomerular microcirculation in the rat. LX-A selectively reduces afferent (R_A), but not efferent (R_E) arteriolar resistance thereby leading to increases in glomerular plasma flow rate (Q_A) and net transcapilary hydraulic pressure difference (ΔP). This results in an increase in single nephron (SN) GFR, an effect partially offset by a simultaneous fall in the glomerular capillary ultrafiltration coefficient (K_f). (From Ref. 6, with permission.)

REFERENCES

1. C. B. Wilson and F. J. Dixon, Renal response to immunological injury, in "The Kidney," B. M. Brenner and F. C. Rector Jr., eds. Saunders, Phila. (1986).

2. K. F. Badr, C. Baylis, J. M. Pfeffer, M. A. Pfeffer, R. J. Soberman, R. A. Lewis, K. F. Austen, E. J. Corey, and B. M. Brenner, Renal and systemic hemodynamic responses to intravenous infusion of leukotriene C_4 in the rat, Circ. Res., 54:492, (1984).

3. K. F. Badr, B. M. Brenner, and I. Ichikawa. Effects of leukotriene D_4 on glomerular dynamics in the rat. Am. J. Physiol. In Press.

4. K. F. Badr, V. E. Kelley, H. G. Rennke, and B. M. Brenner, Roles for thromboxane A_2 and leukotrienes in endotoxin-induced acute renal failure, Kidney Int., 30:474, (1986).

5. K. F. Badr, A. Gung, G. F. Schreiner, M. Wasserman, and I. Ichikawa, Reversal of antiglomerular basement membrane antibody-induced fall in the glomerular ultrafiltration coefficient by the leukotriene D_4 antagonist SK&F 104353. Kidney Int., 31:363, (1987) (Abstr.).

6. K. F. Badr, C. N. Serhan, K. C. Nicolaou, and B. Samuelsson, The action of lipoxin-A on glomerular microcirculatory dynamics in the rat, Biochem. Biophys. Res. Commun., In Press.

7. C. N. Serhan, M. Hamberg, and B. Samuelsson, Lipoxins: Novel series of biologically active compounds formed from arachidonic acid in human leukocytes, Proc. Nat'l Acad. Sci. U.S.A., 81:5335, (1984).

8. C. N. Serhan, K. C. Nicolaou, S. E. Webber, C. A. Veale, S-E Dahlen, T. J. Puustinen, and B. Samuelsson, Lipoxin A: Stereochemistry and biosynthesis. J. Biol. Chem., 261:16340 (1986).

9. K. C. Nicolaou, C. A. Veale, S. E. Webber, and H. Katerinopoulos, Stereocontrolled total synthesis of lipoxins A, J. Amer. Chem. Soc., 107:7515, (1985).

10. I. Ichikawa, D. A. Maddox, M. G. Cogan, and B. M. Brenner, Dynamics of glomerular ultrafiltration in euvolemic Munich-Wistar rats, Renal Physiol., 1:121, (1978).

11. W. M. Deen, J. L. Troy, C. R. Robertson, and B. M. Brenner, Dynamics of glomerular ultrafiltration in the rat. IV. Determination of the ultrafiltration coefficient, J. Clin. Invest., 52:1500, (1973).

12. G. G. Vurek, and S. E. Pegram, Fluorometric method for the determination of nanogram quantities of inulin, Anal. Biochem., 16:409, (1966).

13. J. Fuhr, J. Kazmaczyk, and C. D. Kruttgen, Eine einfache colorimetrische Methode zur Inulinbestimmung fur Nieren-Clearanceuntersuchungen bei Stoffwechselgesunden und Diabetikern, Klin. Wochhenschr., 33:729, (1955).

14. J. W. Viets, W. M. Deen, J. L. Troy, and B. M. Brenner, Determination of serum protein concentration in nanoliter blood samples using fluorescamine or o-phthaldehyde, Anal. Biochem., 88:513, (1978).

15. A. Rosenthal, and C. R. Pace-Asciak, Potent vasoconstriction of the isolated perfused kidney by leukotrienes C_4 and D_4, Can. J. Pharmacol., 61:325, (1983).

16. R. Barnett, P. Goldwasser, L. A. Scharschmidt, and D. Schlondorff, Effects of leukotrienes on isolated rat glomeruli and cultured mesangial cells, Amer. Jour. Physiol., 19:F838, (1986).

17. M. Simonson and M. J. Dunn, Leukotriene C_4 and D_4 contract rat glomerular mesangial cells, Kidney Intern., 30:524, (1986).

18. R.Ardaillou, L. Baud, and J. Sraer, Leukotrienes and other lipoxygenase products of arachidonic acid synthesized in the kidney, Amer. J. Med., 81:(2B):12, (1986).

19. R. J. Soberman, T. W. Harper, D. Betteridge, R. A. Lewis, and K. F. Austen. Characterization and separation of the arachidonic acid 5-lipoxygenase and linoleic acid w-6 lipoxygenase (arachidonic acid 15-lipoxygenase) of human polymorphonuclear leukocytes, J. Biol. Chem., 260:450, (1985).

20. J. Y. Vanderhoek, M. T. Karmin, and S. L. Ekborg, Endogenous hydroxyeicosatetraenoic acids stimulate the human polymorphonuclear leukocyte 15-lipoxygenase pathway, J. Biol. Chem., 260:15482, (1985).

11

EFFECTS OF LIPOXINS A AND B ON FUNCTIONAL RESPONSES OF HUMAN GRANULOCYTES

Jan Palmblad, Hans Gyllenhammar, and Bo Ringertz

Dept. of Medicine 3, Karolinska Institutet, Södersjukhuset
S-100 64 Stockholm
Sweden

ABSTRACT

The effects of synthetic lipoxin A and B (LXA and LXB) as well as several of their isomers were assessed as inducers of functional responses of human granulocytes. LXA stimulated migration at 1 nM and exhibited a variable effect as inducer of chemiluminescence at concentrations $\geq$ 1 µM. Both migration and chemiluminescence evoked by LXA proved to be highly stereospecific, since e.g. 6S-LXA was less active than LXA. Neither synthetic LXA nor several of its stereoisomers provoked degranulation, aggregation, membrane potential changes or intracellular calcium fluxes. In addition, LXB and its isomers did not stimulate aggregation, degranulation or chemiluminescence. Pretreatment of granulocytes with LXA did not modify subsequent challenges with leukotriene B_4. Together, these results indicate that granulocyte responses to LXA are highly stereospecific and are mediated by other mechanisms than those evoked by leukotriene B_4.

INTRODUCTION

Leukotriene B_4 (LTB_4) plays an important role for granulocyte physiology[1]. LTB_4 is, at least, of similar potency as bacterial formylpeptides (e.g. fMLP) and complement split products (e.g. C5a) as inducer of granulocyte chemotaxis, adherence and aggregation [2-5]. The efficacy of LTB_4 as a promotor of degranulation was approximately half that of fMLP at equimolar concentrations [6,7], whereas the oxidative metabolism was less activated, approximately half to one tenth of that of fMLP [8,9]. These effects endow LTB4 with a unique profile of activity.

Recently, a new series of lipoxygenation products from ionophore A23187 stimulated granulocytes were reported and termed lipoxins [10]. The biochemistry of these substances is detailed elsewhere in this volume. The two major compounds of this series are lipoxin A (LXA) and lipoxin B (LXB). Both LXA and LXB were reported to possess discrete biological activities. LXA, obtained from incubations of leukocytes, was observed to cause secretion of superoxide anion and elastase from human granulocytes [10], contraction of lung strips [11] and inhibition of the cytotoxic activity of human NK cells [12]. In each of these systems it appeared that the biological activities of LXA and LXB differed from those of either leukotrienes or prostaglandins.

Recently, the complete stereochemistry of LXA and LXB and several of their isomers have been determined [13,14]. In addition to LXA, 11-trans-LXA, 6S-LXA and 6S-11-trans-LXA has been identified. Also, synthetic lipoxins have been generated [13,14]. When added to human NK cells [15], guinea pig lung strips or hamster cheek pouch [11,13], synthetic LXA exhibits biological activities similar to those observed with the biologically derived material.

This report concerns some effects of synthetic lipoxins on several human granulocyte functional responses. In addition to lipoxin B 3 of its isomers, (8-trans-, 14S-8-trans- and 14S-LXB) were examined.

MATERIAL AND METHODS

Eicosanoids. Synthetic lipoxins were prepared as in [16,17]. Lipoxins were used as free acids or methyl esters, as indicated. In some experiments LXA was isolated from suspensions of mixed human leukocytes (i.e. neutrophils, eosinophils, basophils etc.)[10,13]. - Synthetic LTB_4 was obtained from BioMol Research Lab. (Philadelphia, PA). LTB_4 samples were handled as described for lipoxins.

Chemicals. Percoll and dextran T500 were from Pharmacia (Upsala, Sweden), A23187 from Calbiochem (La Jolla, CA), fMLP from Peninsula Lab. (San Carlos, CA) and Hanks´ balanced salt solution (HBSS) from Natl. Bacteriol. Lab. (Stockholm, Sweden). All other reagents were from Sigma Chem. Co. (St. Louis, MO) and of highest obtainable purity.

Cell isolation. Heparinized blood samples were obtained from healthy members of the staff. None were on medication. Granulocytes were isolated by a one-step Percoll technique [5]. With this technique, platelets were removed by an initial centrifugation step, and neutrophils comprised >95 % of the cells, as determined from stained smears. Platelets represented approximately one in ten granulocytes [5]. Eosinophil and basophil counts were not regularly checked. In some experiments mixed leukocyte suspensions were obtained by sedimentation of erythrocytes with dextran [2]. Cells were suspended in HBSS at pH 7.45 and kept at +4°C until use, followed by 15 min. incubation at 37°C. In some experiments, as indicated, cells were treated with cytochalasin B (5µg/ml) for 3 min. at +37°C.

Functional assays

Granulocyte chemotaxis was measured with a Boyden chamber technique[18], (Neuroprobe Inc., Bethesda, MD). Granulocytes were allowed to migrate into 3 µm pore size cellulose nitrate filters (Microfiltration Systems, Dublin, CA) for 45 min. at 37°. HBSS alone, fMLP and LTB_4 were used as standard chemotactic stimuli. Migration was expressed as the mean distance migrated (in µm) by the leading cells and as the mean numbers of cells per microscopic field, at a depth of 50 µm into the filters. Results are expressed as net migration or cell numbers, i.e. stimulated migration minus spontaneous migration.

The oxidative metabolism of granulocytes was assessed both as luminol--enhanced chemiluminescence (LCL) [8], and as ferricytochrome C reduction [9]. Briefly, for LCL assays purified granulocytes or dextran sedimented leukocytes were mixed with luminol (0.17 mM).

The pH was kept at 7.6-8, but in some experiments other pH were obtained by adding NaOH or HCl to the buffer before addition of cells. Chemiluminescence was followed continously (at room temperature) with a Luminometer 1 50 (LKB, Bromma, Sweden) [8]. Superoxide ion production was analyzed by the superoxide dismutase inhibitable cytochrome C reduction method[9,10] at 37° with continuous stirring of cytochalasin treated granulocytes. Absorbance was read at 550 nm and superoxide ion production was calculated as nmol reduced cytochrome C using an absorption coefficient of 21.1 mM^{-1} cm^{-1}.

Granulocyte aggregation was measured in a standard platelet aggregometer (Model 300 BD, Payton Associates, Buffalo, NY)[5].

Elastase release from cytochalasin B treated granulocytes was monitored as change of absorbance of N-t-Boc-L-Ala-pNP (10 µM) at 360-390 mm [9]. All measurements were performed at 37°C with continuous stirring

Shape changes were followed by interference contrast microscopy, as described [19].

Membran potential changes were followed as the change of fluorescence at 510 nm of the cyanine dye di-0-C5(3) at 20 nM after excitation at 460 nm, as described [20]. Measurements were made in a Perkin-Elmer spectrofluorometer, at 37°C and with continous stirring of cell suspensions.

Intracellular calcium concentrations were calculated from the change of Fura-2AM fluorescence [20]. After loading cells with 0.5 μM Fura-2AM, washing, fluorescence was excited at 340 nm and recorded at 510 nm. The system was calibrated with EGTA, Tris-buffer, Triton X-100 and CaCl₂ as described [20].

All assessments of neutrophil functions were performed within 4 hours after cell separation. In each experiment LTB₄, fMLP and appropriate solvents were run for comparison with test compounds. Included in this study are only experiments where these controls were within the previously established range of activity.

Statistical methods. Student's t-test was used for analyses.

RESULTS

When purified granulocytes were exposed to synthetic lipoxins as chemoattractants, stimulated migration, assessed both as leading front cell distance and number of cells at 50 μm, was observed for LXA free acid with a peak effect at 1 nM (p< 0.001 compared with migration against HBSS)(Fig. 1). Neither the methyl ester of LXA nor 6S-LXA, 6S-11-trans-LXA or 11-trans-LXA conferred significantly stimulated migration (Fig. 1). A significant effect was seen with 14S-8-trans-LXB at 0.1 μM (p<0.01) but not, with LXB or 14S-LXB. However, LTB₄ and fMLP induced a considerably more pronounced migration (Fig. 1). Biologically derived LXA methyl ester and LXB-free acid were also active (Fig. 1).

When chemiluminescence was used as a means to assess oxidative metabolism a complex response pattern evolved. Whereas most experiments (n=11) did not disclose any luminescent response of cytochalasin treated granulocytes to lipoxins (10 μM or less) but readily to LTB₄ and fMLP (0.1 μM and less), three donor cell preparations exhibited a clear response to synthetic LXA. For these experiments the response to LXA proved to be stereospecific. Here, the light emission induced by the free acid of LXA at 10 μM was 19.5 mV (n=1), at 1 μM 10 ± 8 mV (n=3) but at 0.1 μM none (n=3) (Fig. 2). Moreover, those granulocyte preparations that reacted to lipoxins did so also when cells had not been treated with cytochalasin B, although the net light emission was then only 1/10 of that of treated cells. That difference of LCL is similar to what was observed with LTB₄ and fMLP. The kinetics of the LXA induced LCL response from untreated granulocytes was considerably slower than that of LTB₄, rather being more alike the fMLP response (Fig. 2). With cytochalasin treatment such differences of response kinetics disappeared. Moreover, 6S-LXA, and less so LXA methyl ester, elicited LCL, being less potent than LXA free acid (Fig. 2). In all these experiments LTB₄ and fMLP responses appeared normal, and HBSS and ethanol containing controls did not produce any LCL. When LXA free acid was added to cell free, but luminol containing samples, no LCL ensued.

In order to determine whether other constituents of cell preparations interacted with granulocytes to produce LCL several experiments were performed with unpurified dextran sedimented and cytochalasin B treated leukocytes and platelets exposed to LXA. Neither that measure, nor changing the pH of purified granulocyte suspensions between 7 and 9, turned unreactive samples into LCL positive.

Superoxide ion production from cytochalasin treated purified granulocytes was also negative for LXA free acid, whereas LTB₄ and fMLP were positive (n=3). Because of a lack of cells, LCL positive samples were not tested in parallel for superoxide ion generation. It should be noted that

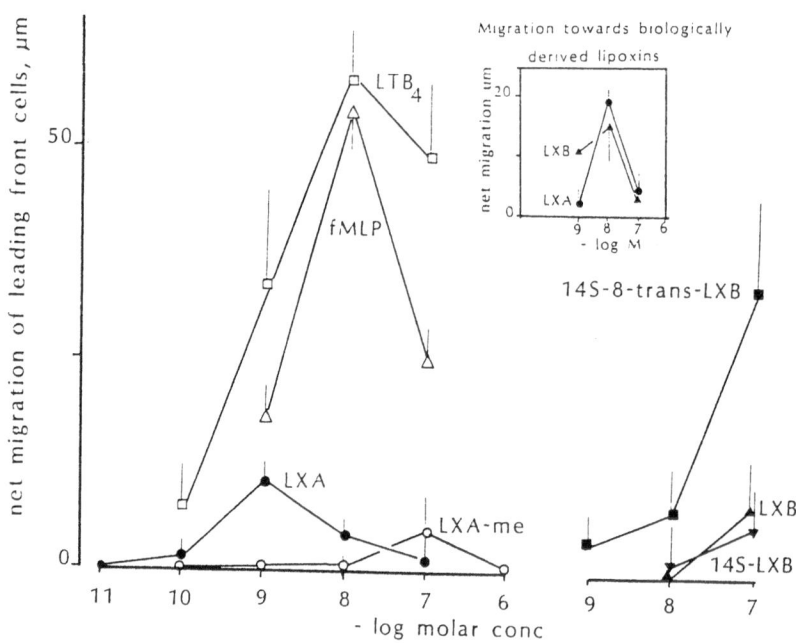

Fig. 1. Migration of granulocytes in modified Boyden chambers in response to synthetic lipoxins, fMLP and LTB_4. The results are given as the distance to the leading front cells and expressed as net migration, i.e. stimulated minus spontaneous migration (to HBSS), the latter being 41 ± 2 nm (n=20). Each mean value is based on 4-19 separate experiments; more specifically, for peak migration values in response to fMLP (△) and LTB_4 (□) n was 18, for LXA (●) n=11, for LXA methyl ester (LXA-me, 0) n=8, for LXB (▲) n=3, for 14S-8trans-LXB (■) n=4 and for 14S-LXB (▼) n=4. The number of granulocytes having migrated 50 µm into the cellulose nitrate filters was for HBSS 1.3 ± 0.7 per high power magnification field, for 10 nM of LTB_4 and fMLP the net number of cells (i.e. values for spontaneous migration have been subtracted) 49 ± 10 and 69 ± 13, respectively, for 1 nM of LXA 4.2 ± 2.6, for 100 nM of LXA-me2.7 ± 2.5, for LXB and 14S-LXB (both at 100 nM) none, and, for 14S-8trans-LXB, also at 100 nM, 30 ± 10 cells (means ± SE values). Not shown in this figure are the values for 6S-LXA, which never exceeded spontaneous migration when assessed between 0.01 and 100 nM (n=3). Likewise, no response was noted for 6-trans isomers of LXA.

responses to LTB_4 were easier discernable with the LCL than the cytochrome C assay, since LCL responses of granulocytes to even 5 nM could be detected in the absence of cytochalasin treatment of cells, conditions which did not confer any measurable cytochrome C reduction.

Next, when granulocyte aggregation was assessed a small aggregating response could be observed to 10 µM of 6S-LXA, whereas LXA did not evoke such a response (Table 1).

In addition, no obvious shape changes could be seen by light microscopy in response to 1 µM LXA. Elastase release membrane potential changes and alterations of intracellular calcium changes were not evident from granulocytes exposed to synthetic LXA free acid at 10 µM (n=4)(Fig. 3, 4 & 5). In contrast, fMLP and LTB_4 conferred responses in these assays, the latter causing more rapidly emerging but also more rapidly disappering than fMLP.

We next assessed whether prior exposure of purified granulocytes to
LXA would modify their responses to subsequent challenge by either LTB$_4$
or fMLP. In these experiments LXA was added 100 sec before a stimulation
with LTB$_4$ or fMLP. This hypothesis was tested in the LCL assay. It was
found that those cytochalasin treated granulocytes, which did not react to
LXA with LCL, responded normally to 0.1 μM LTB$_4$, i.e. 102% of the LCL that
those cells did, which had been treated only with the corresponding solvent
(ethanol) concentration (Table 1). In contrast, cells which did respond
with LCL to LXA (at 10 or 1 μM) produced less LCL to LTB$_4$ (Table 2). When
fMLP (at 0.1 μM) was used as the subsequent stimulus an enhanced response
was noted, being appr. 124% of controls.

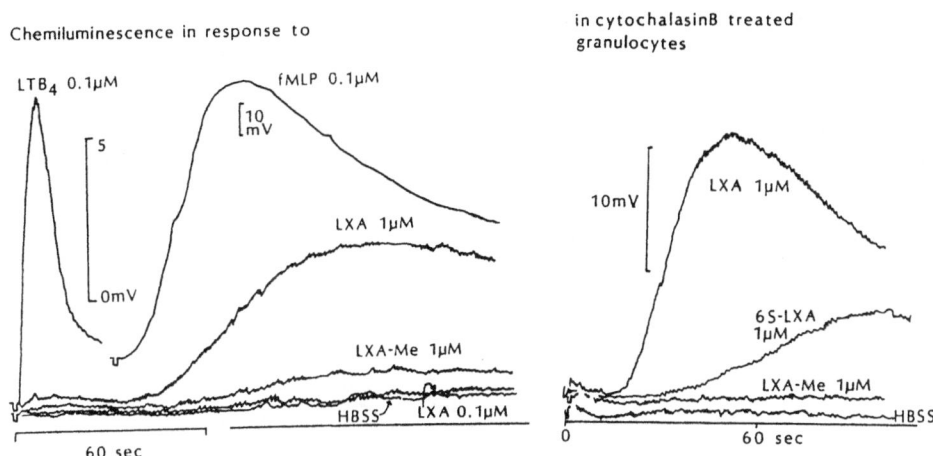

Fig. 2. A. Chemiluminescence (CL) of granulocytes stimulated with lipoxin
A, fMLP and LTB$_4$. The figure represents actual tracings from one of the
experiments where granulocytes not treated with cytochalasin B responded.
The 5mV scale applies to all tracings except the fMLP response, where a
separate 10 mV scale is inserted. B. Chemiluminescence of cytochalasin B
treated granulocytes from another experiment, where fMLP and LTB$_4$ also conferred typical CL responses.

Table 1
Granulocyte aggregation. Mean and SE values.
Stimulant Concentration Max. aggregation

	μM	mm
HBSS		1 ± 1
LXA	10	9 ± 9
-"-	1	2 ± 2
6S-LXA	10	33 ± 1
-"-	1	16 ± 13
LTB$_4$	0.1	30 ± 6

Granulocytes were treated with cytochalasin B, as described, and subsequently exposed to the stimulants. The ensuing maximal deflection from
the baseline, measured in mm, was used for calculations of aggregation.

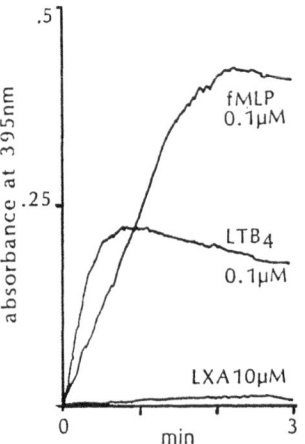

Fig. 3. Elastase release (analysed as detailed in the Methods section) in response to fMLP, LTB$_4$ and LXA. The figure represents one actual and representative tracing out of four separate experiments, with identical results.

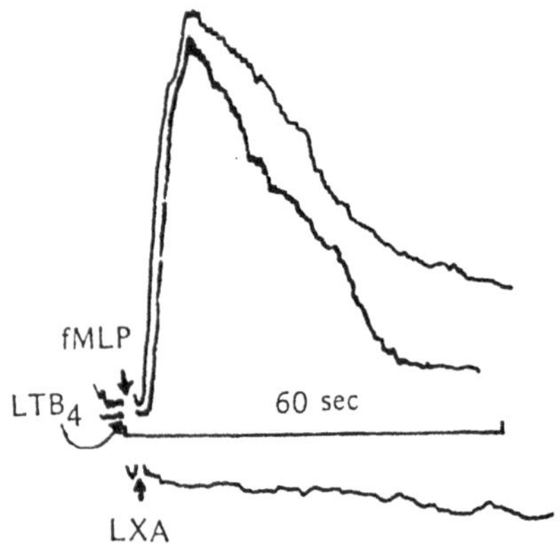

Fig. 4. Intracellular calcium concentration changes, assessed by Fura-2AM fluorescence. The figure represents one of 4 separate experiments with identical results.

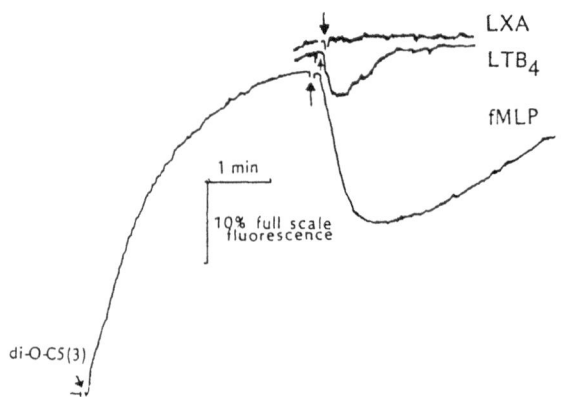

Fig. 5. Membrane potential changes, assessed by di-O-C5(3) fluorescence. The figure represents one of 3 separate experiments with identical results.

DISCUSSION

It has been shown that effects of lipoxins appear to be induced by mechanisms that are distinct from those of either leukotrienes or prostaglandins [11]. Thus, LTB$_4$ causes no activation of purified human placental protein kinase C or its translocation from neutrophil cytosol to membranes [21, 22], whereas LXA was recently shown to cause protein kinase activation in isolated preparations of the enzyme[21].

The fact that synthetic LXA induced chemotaxis (Fig. 1) and oxidative response (Fig. 2) demonstrate that lipoxin A may be of significance for granulocyte physiology.

There are a number of characteristic features for migratory response

142

Table 2
Modulation of granulocyte LCL responses by LXA

Cell reaction to LXA:[1]	% response to second challenge with [2]	
	LTB$_4$	fMLP
positive response	65 $\pm$ 15 (3)	n.d.
negative response	102 $\pm$ 22[3] (6)	124 $\pm$ 18 (4)

Figures within brackets denote number of separate experiments. n.d. = not determined. Results are given as means $\pm$ SE values for peak LCL values.
1/ Purified granulocytes were first treated with cytochalasin B, then challenged with LXA. Depending on whether cells exhibited LCL to any of the used concentrations (10 or 1 μM) experiments were grouped either as positive or negative responses.
2/ LTB$_4$, or fMLP, both at 0.1 μM, were added 100 sec after exposure of granulocytes to 1 μM LXA. The results are given as per cent of cells also challenged with either fMLP or LTB$_4$ but where granulocytes first had been exposed only to ethanol (being the solvent for LXA), 0.1-0.01 %. Ethanol treatment reduced responsiveness to appr. 80 % of HBSS treatment alone.
3/ When LXA was used at 0.1 μM for pretreatment the subsequent LTB$_4$ response was 98 $\pm$ 37 % (n=3).

to LXA. Firstly, at the optimal LXA concentration, 1 nM, which is one tenth of that of LTB$_4$ and fMLP, the distance migrated by leading front cells was considerably shorter than for LTB$_4$ or fMLP. This suggests that LXA may cause migration in a subset of granulocytes or alternatively, that LXA may be rapidly transformed by granulocytes so that no stable chemotactic gradient is established. Secondly, a definite structure activity relationship for lipoxins was established. Both synthetic and biologically derived LXA (but not 6S-LXA or the two transisomers) were effective as chemoattractants (Figure 1). In addition, the methyl ester of LXA showed little to no effect. These findings indicate stereospecific determinates for induction of chemotaxis. Thirdly, the finding of a discrete LXA concentration at which migration was stimulated (whereas both higher and lower concentrations failed to do so) is in agreement with what is observed for LTB$_4$ and fMLP and other stimuli believed to exert their chemotactic effect by means of high affinity surface receptors [23,24]. This finding may suggest the presence and active expression of cellular receptors for the migratory response to LXA.

At present, we have no explanation as to why only cells from certain donors reacted to LXA with respect to LCL. This dicotomy into responsive prone or unresponsive granulocytes appeared not to be due to selection or exclusion of leukocytes or platelets during the cell purification process, since unpurified dextran sedimented cell preparations were also unresponsive. Further work is under way to clarify this issue. Nonetheless, as shown here for migration, there was a strict structure-activity relationship in the reactive preparations which favored LXA as the active compound (Figure 2). The finding that the methyl ester of LXA was ineffective is in agreement with findings concerning LTB$_4$ and its methyl ester, where the latter did not evoke an oxidative response in granulocytes [25].

The LCL assay offers a unique possibility to follow continously the stimulus-response coupling of granulocytes to a certain stimulus. LTB_4 initiates a very fast response, which also subsides rapidly (Figure 2). For fMLP, activation is apparently slower in onset than that of LTB_4. Also the termination of the response to fMLP is again slower than LTB_4. LXA initiates an even more delayed response than either LTB_4 or fMLP. In this particular respect the kinetics of LXA responses resemble LCL and aggregatory responses induced by either ionophore A23187 or phorbol myristate acetate stimulated granulocytes (Palmblad et al, to be published).

Most cell preparations exposed first to LXA and subsequently challenged with LTB_4 reacted to the latter without a change in responsiveness to the second challenger. This phenomenon might be explained by separate activation mechanisms for these two stimuli. However, LXA pretreatment of granulocytes confered an enhanced responsiveness to fMLP. This observation is in agreement with other reports, where pretreatment with one stimulus renders granulocytes primed to a subsequent challange by unrelated stimulus [26]. Taken together, the results of these experiments suggest that LXA induces granulocytes responses by means of a discrete mechanism which appears to differ from those of other stimuli. In addition, they suggest that granulocyte responses to fMLP can be primed by prior exposure of the cells to LXA.

The here presented results have partly been published separately [27].

Acknowledgements. Drs. KC Nicolaou and CN Serhan kindly provided the lipoxins. This study was supported by grants from the Swedish Medical Research Council (19P-7095, 19X-5991, 19I-05991), the Funds of P. & A. Hedlund, C. Bergh, Förenade Liv Mutual Insurance Company (HG,BR), KabiVitrum AB, Karolinska institute, Södersjukhuset and the Swedish Medical Society. The skilled technical assistance of Mrs I Friberg, S Riddez and Mr B Cottell is gratefully acknowledged.

REFERENCES

1. Samuelsson B: Leukotrienes. Science 220:568, 1983.
2. Palmblad J, Malmsten CL, Udén A-M, Rådmark O, Engstedt L, Samuelsson B: Leukotriene B_4 is a potent and stereospecific stimulator of neutrophil chemotaxis and adherence. Blood 44:37, 1981.
3. Goetzl EJ, Pickett WC: The human PMN leukocyte chemotactic activity of complex hydroxy-eicosatetraenoic acids (HETEs). J Immunol 125:1789, 1980.
4. Dahlén SE, Björk J, Hedqvist P, Arfors KE, Hammarström S, Lindgren JÅ, Samuelsson B: Leukotrienes promote plasma leakage and leukocyte adhesion in postcapillary venules. Proc Natl Acad Sci USA 78:3887, 1981.
5. Ringertz B, Palmblad J, Rådmark O, Malmsten CL: Leukotriene induced neutrophil aggregation in vitro. FEBS Lett 147:180, 1982.
6. Hafström I, Palmblad J, Malmsten CL, Rådmark O, Samuelsson B: Leukotriene B_4 - a stereospecific stimulator for release of lysosomal enzymes from neutrophils. FEBS Lett 130:146, 1981.
7. Feinmark S, Lindgren JÅ, Claesson HE, Malmsten CL, Samuelsson B: Stimulation of human leukocyte degranulation by leukotriene B_4 and its ω-oxidized metabolites. FEBS Lett 136:141, 1981.
8. Palmblad J, Gyllenhammar H, Lindgren JÅ, Malmsten CL: Effects of leukotrienes and f-Met-Leu-Phe on oxidative metabolism of neutrophils and eosinophils. J Immunol 132:3041, 1984.
9. Serhan CN, Radin A, Smolen JE, Korchak HM, Samuelsson B, Weissman G: Leukotriene B_4 is a complete secretagogue in human neutrophils. A kinetic analysis. Biochem Biophys Res Commun 107:1006, 1982.
10. Serhan CN, Hamberg M, Samuelsson B. Lipoxins: novel series of biologically active compounds formed from arachidonic acid in human leukocytes. Proc Natl Acad Sci USA 81: 5335-5339, 1984.

11. Dahlén SE, Raud J, Serhan CN, Björk J, Samuelsson B: Biological activities of lipoxin A includes lung strip contraction and dilation of arterioles in vivo. Acta Physiol Scand 1987, in press.

12. Ramstedt U, Ng J, Wigzell H, Serhan CN, Samuelsson B: Action of novel eicosanoids lipoxin A and B on human natural killer cell cytotoxicity: effects on intracellular cAMP and target cell binding. J Immunol 135: 3434, 1985.

13. Serhan CN, Nicolaou KC, Webber SE, Veale CA, Dahlén SE, Puustinen TJ, Samuelsson B: Lipoxin A: sterochemistry and biosynthesis. J Biol Chem 261:16340, 1986.

14. Serhan CN, Hamberg M, Samuelsson B, Morris J, Wishka DG: On the stereochemistry and biosynthesis of lipoxin B. Proc Natl Acad Sci USA 83: 1983, 1986.

15. Ramstedt U, Serhan CN, Nicolaou KC, Webber SE, Wigzell H, Samuelsson B: Lipoxin A induced inhibition of human natural killer cell cytotoxicity: Studies on stereospecificity of inhibition and mode of action. J Immunol 138:266, 1987.

16. Nicolaou KC, Veale CA, Webber SE, Katerinopoulos HK: Stereocontrolled total synthesis of lipoxin A. J Amer Chem Soc 107:7515, 1985.

17. Nicolaou KC, Webber SE: Stereocontrolled total synthesis of lipoxins B. Int J Methods Synth Org Chem 6:453, 1986.

18. Gyllenhammar H, Ringertz B, Becker W, Svensson J, Palmblad J: Essential fatty acid deficiency in rats. Immunol Letters 13:185, 1986.

19. Smith CW, Hollers JC, Patrich RA, Hassett C: Motility and adhesiveness in human neutrophils. J Clin Invest 63:221, 1979.

20. Metcalf JA, Gallin JI, Nauseef WM, Root RK: Laboratory manuel of neutrophil function. Raven Press, New York, 1986.

21. Hansson A, Serhan CN, Haeggström J, Ingelman-Sundberg M, Samuelsson B: Activation of protein kinase C by lipoxin A and other eicosanoids. Biochem Biophys Res Commun 134:1215, 1986.

22. Nishihira J, McPhail LC, O'Flaherty JT: Stimulus-dependent mobilization of protein kinase C. Biochem Biophys Res Commun 134:587, 1986.

23. Seligmann BE, Fletcher MP, Gallin JI: Adaptation of human neutrophil responsiveness to the chemoattractant N-formylmethionylleucylphenylalanine. J Biol Chem 257:6280, 1982.

24. Goldman DW, Goetzl EJ: Heterogeneity of human polymorphonuclear leukocyte receptors for leukotriene B_4. J Exp Med 159:1027, 1984.

25. Serhan CN, Hamberg M, Samuelsson B: Novel mechanisms in the arachidonic acid cascade: formation of lipoxins. In Adv Inflammation Res (ed F Russo-Marie et al). Raven Press, New York 10:117, 1985.

26. Gay JC, Beckman JK, Brash AR, Oates JA, Lukens JN: Enhancement of chemotactic factor-stimulated neutrophil oxidative metabolism by leukotriene B_4. Blood 64:780, 1984.

27. Palmblad J, Gyllenhammar H, Ringertz B, Serhan CB, Samuelsson B: The effects of lipoxin A and lipoxin B on functional responses of human granulocytes. Biochem Biophys Res Commun, in press.

12

LIPOXINS OF THE 5-SERIES DERIVED FROM EICOSAPENTAENOIC ACID

Bernd W. Spur, Crawford Jacques, Attilio E. Crea, and Tak H. Lee

Department of Medicine, UMDS
Guy's Hospital, London
SE1, UK

INTRODUCTION

Arachidonic acid released from membrane phospholipids during cell activation is metabolised by the cyclo-oxygenase pathway to prostaglandins and thromboxanes and by the 5-lipoxygenase pathway to leukotrienes. In the 5-lipoxygenase pathway, arachidonic acid is metabolised through two sequential intermediates, 5-hydroperoxy eicosatetraenoic acid (5-HPETE) and leukotriene A4 (LTA4), to leukotriene B4 (LTB4) and to leukotriene C4 (LTC4) and its conversion products, leukotriene D4 (LTD4) and leukotriene E4 (LTE4).

Eicosapentaenoic acid (EPA) and docosahexaenoic acid (DCHA) termed N-3-fatty acids to reflect the position of the double bond furthest from the carboxylic acid, are each prominent in fish-oil-enriched diets. Both these marine fatty acids competitively inhibit the utilization of arachidonic acid by the cyclo-oxygenase. The prostaglandin endoperoxide and thromboxane A3 derived from EPA have attenuated platelet-aggregating activity as compared with the arachidonic acid derived products. When compared to arachidonic acid, EPA is a preferred substrate for product generation by way of the 5-lipoxygenase in subcellular fractions of human neutrophils. LTB5 is 30- to 60- fold less potent than LTB4 in eliciting neutrophil chemotaxis. In addition, it has been shown that EPA and its oxygenated product, LTA5, inhibit LTB4 formation at the level of the LTA4 hydrolase. Relatively little is known about EPA metabolism via the 15-lipoxygenase pathway. Recently Wong has demonstrated that 15-hydroperoxy eicosapentaenoic acid (15-HPEPE) can be converted to a novel series of the hydroxy pentaenes, namely lipoxin A5 (LxA5) and lipoxin B5 (LxB5) (1) (Fig 1). Because of the potential importance of dietary enrichment with fish oil lipids in health and disease (2), we were prompted synthesize these newly discovered metabolites and to study their biological activities.

Since we had reported the first synthesis of the 5-series leukotrienes (3), we were able to complete the first synthesis of LxA5 according to the report by the Merck group (4). Nucleophilic opening of the epoxide ring gave the 5S,6S dihydroxy eicosapentaenoic acid and the 5S, 6R epimer which were separated by SP-HPLC. These compounds were converted separately by 15-lipoxygenase to LxA5 and its 5S, 6S epimer. In these experiments we always observed the presence of the all trans isomers which resulted from isomerisation during the work up and on the surface of the material contained in the HPLC column. This isomerisation could be minimised by the use of an alkaline system for the isolation of these compounds. By utilising LTA5 as the starting substrate, the predominant compound is the 5S, 6S, 15S LxA5 and the minor component is

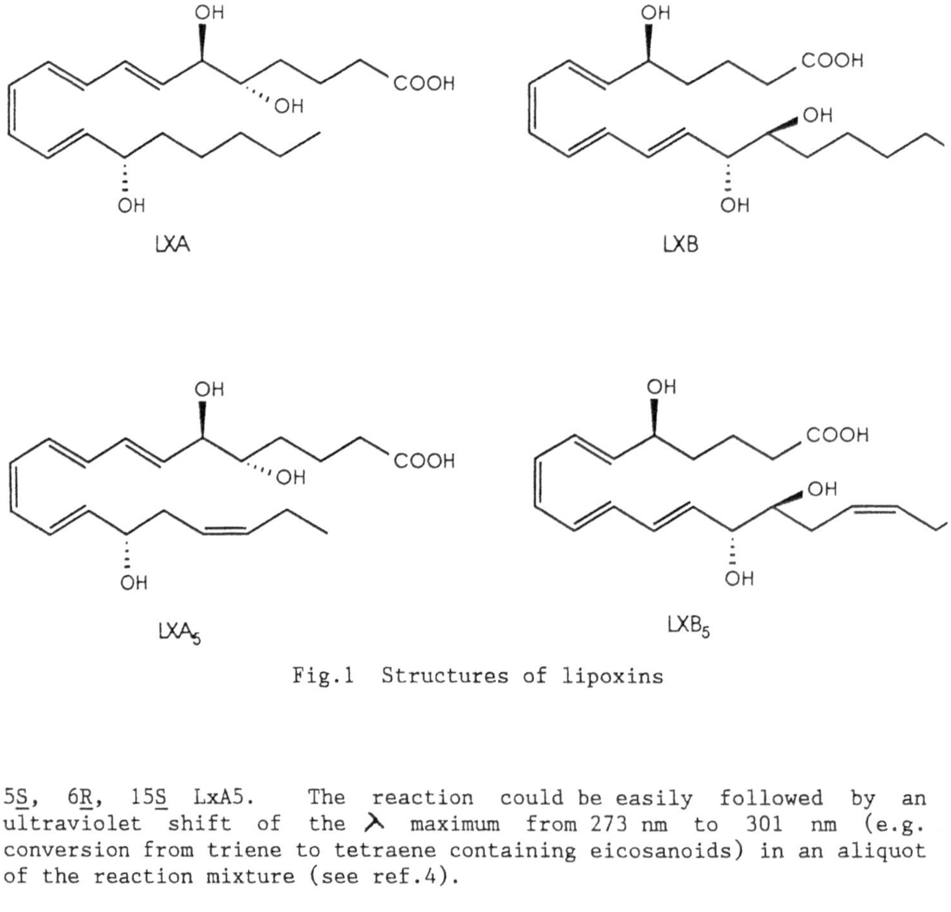

Fig.1 Structures of lipoxins

5S, 6R, 15S LxA5. The reaction could be easily followed by an ultraviolet shift of the λ maximum from 273 nm to 301 nm (e.g. conversion from triene to tetraene containing eicosanoids) in an aliquot of the reaction mixture (see ref.4).

In order to obtain all possible epimers in position 5 and 6 it was necessary to start from the epimeric LTAs (4) and an elegant synthetic route which started from 2-deoxy-d-ribose was developed at Merck (5). The following scheme shows an easy approach to obtaining these epimeric LTAs which is based on the Sharpless epoxidation (Fig.2). This gave the 5S, 6S LTA or, using the enantiochiral director, the double epimer 5R, 6R LTA. Using the same strategy but starting from the cis allylic alcohol the single epimers were obtained.

It was desirable to obtain the lipoxins in the optically active form with a minimum of chromatographic separation steps (6). Starting from 2-deoxy-d-ribose which contained the correct configuration of the 5$\underline{S}$, 6$\underline{R}$ hydroxy group as well as the 14$\underline{R}$, 15$\underline{S}$ centres (7), the key intermediate was obtained with an excellent overall yield. Using the same precursor it was also possible to obtain the 5$\underline{S}$,6$\underline{S}$ epimer as well as the 14$\underline{S}$, 15$\underline{S}$ epimer (7). Stereoselective synthesis of LxA4 and LxB4 has previously been described by Nicolaou (8,9). These compounds can now be obtained from 2-deoxy-d-ribose thereby shortening the synthetic route. Hydrogenation of the double bond in the 17,18 position in an

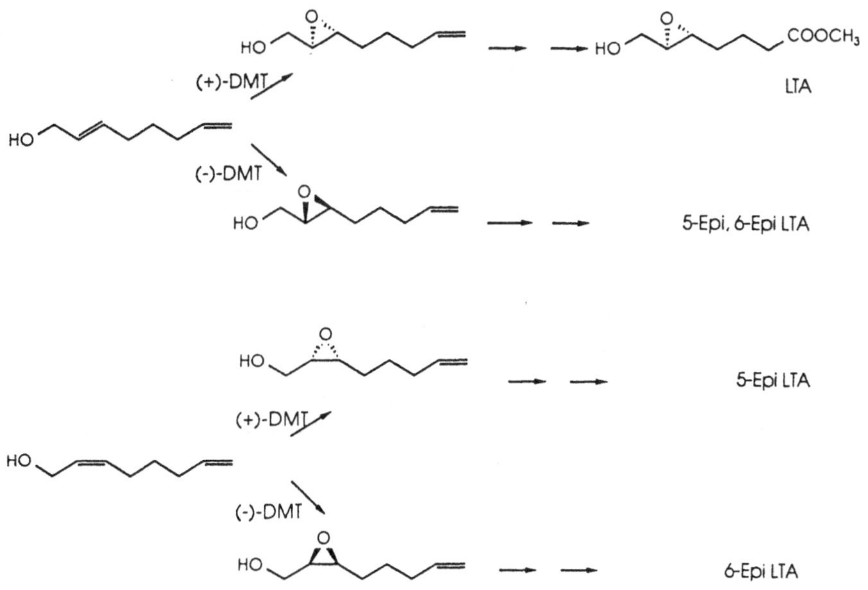

Fig.2 Synthesis of LTA, 5-epi LTA, 6-epi LTA, 5-epi,6-epi LTA

early step but following the same scheme gave LxB4 and the all _trans_ isomers after exposure to iodine. Using the epimerisation it was possible to obtain the 14$\underline{S}$,15$\underline{S}$ epimers which after exposure to iodine resulted in the formation of the all _trans_ LxB5. The critical step was the hydrogenation of the triple bond to the _cis_ double bond. Under a variety of conditions, this reaction resulted in some all _trans_ isomerisation and it must be followed very carefully by HPLC. An elegant synthesis of the biosynthetic intermediate for the formation of LxA4 and LxB4 was recently described by Corey (10) (Fig.3).

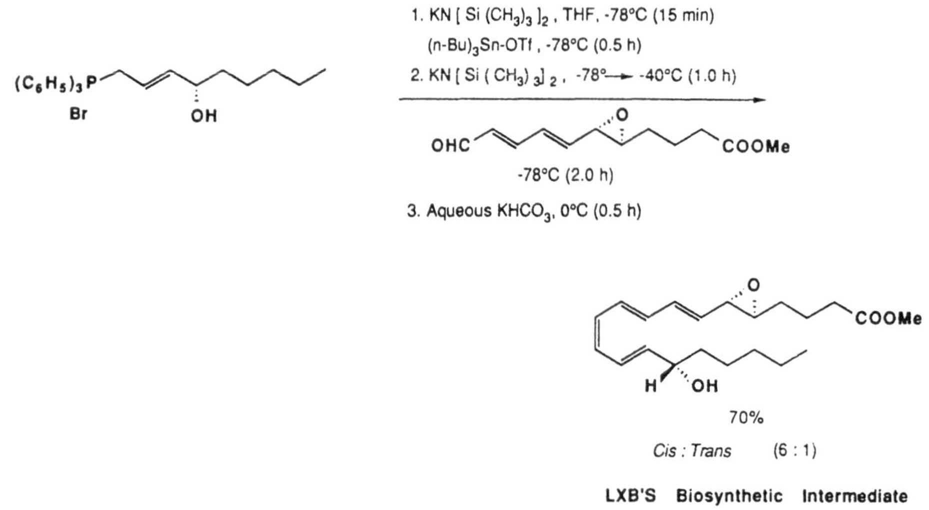

Fig. 3   Corey, Mehrota Synthesis of the Precursor to the Lipoxins
         (Tet.Letters, 1986)

It was not possible to use the same scheme for the synthesis of the proposed biochemical intermediate for lipoxins of the 5- series due to side reactions. We therefore started from D(-)arabinose (11) to construct the 15S segment which then gave the 5S, 6S epoxy 15S hydroxy compound and its all trans isomer in a ratio of nearly 1:1 (Figs.4 & 5).

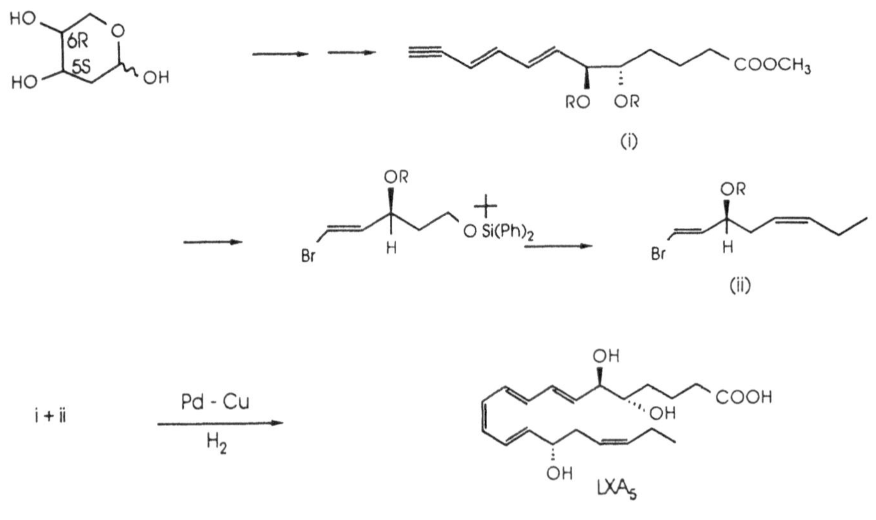

Fig.4   Synthesis of LxA5

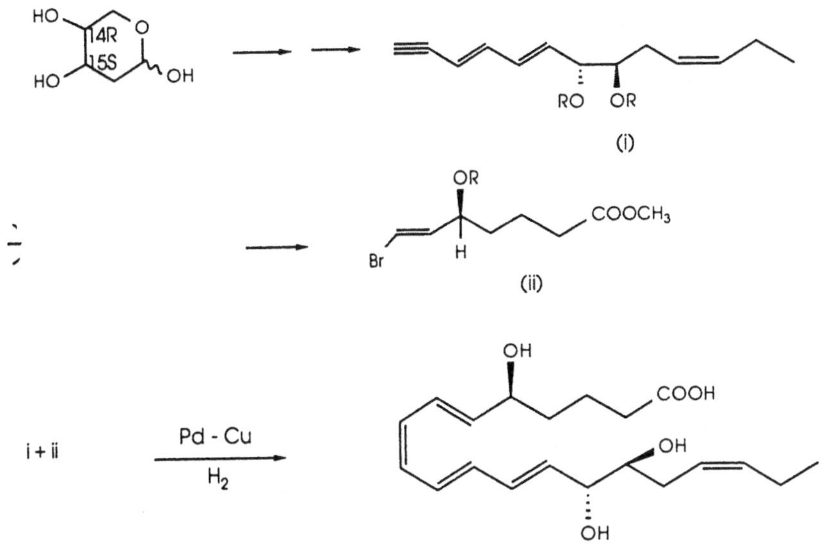

Fig.5 Synthesis of LxB5   R = Si (CH₃)₂ C₄H₉

Lipoxins are ideal candidates for an GC-MS assay utilising an internal deuterated standard. This can be made available by the synthetic scheme outlined below. All the compounds were characterized by 2D-NMR, MS, UV, HPLC and bioassay as well as computer simulated spin analyses (12).

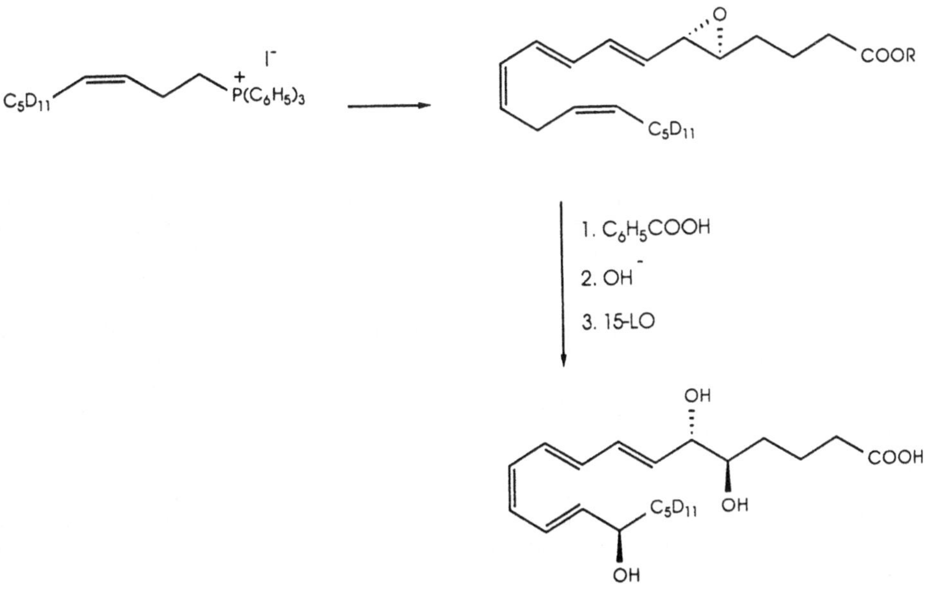

Fig.6 Synthesis of d₁₁ - lipoxin A4

Biological activities of lipoxins

Utilising compounds prepared by total chemical synthesis as outlined above, we compared the biological activities of LxA4 and LxB4 with those of LxA5 and LxB5.

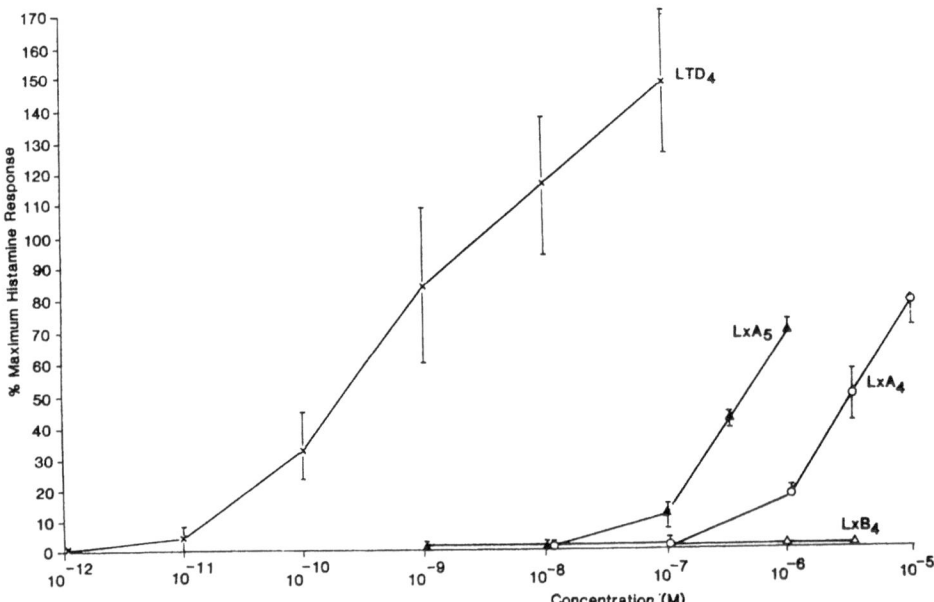

Fig.7 Dose/Response Curves for LTD4, LxA4, LxB4 and LxA5 on Parenchymal Strips

Contractile Activities

The isometric contractile activities of lipoxin A4 and lipoxin B4 for guinea pig lung tissue were evaluated over the concentration range 10^{-8} to 10^{-5} M. LxA4 contracted guinea pig lung parenchymal strips but not tracheal spirals and the concentration eliciting 50% maximum histamine response was 3×10^{-6} M. The LxA4 dose response curve was parallel to that of LTD4 with LxA4 being approximately ten thousand-fold less potent than LTD4. LxA5 was approximately ten-fold more potent than LxA4. The time courses of the contraction elicited by LxA4 and LxA5 were similar to that of LTD4 and they were slow in onset and did not plateau for twenty minutes. The incubation of parenchymal strips with 1 x 10^{-6} M to 3×10^{-5} M FPL 55712 inhibited LxA4 and LxA5 activity in a dose dependent manner. The incubation of tissues with 1 x 10^{-5} M indomethacin did not affect the contractile activity of either LxA4 or LxA5. LxB4 did not constrict parenchymal strips or tracheal spirals (see Fig 7).

152

Cell migration

The ability of lipoxins to enhance neutrophil migration was assessed using a microchemotaxis technique. 5S,6R LxA4 produced a dose dependent enhancement in neutrophil migration. It was approximately 100-fold less potent than LTB4 and in addition only elicited approximately 50% of the maximal response of LTB4. The 5S,6S LxA4 isomer and LxA5 were unable to elicit increased neutrophil migration suggesting that there was a high degree of stereo-selectivity. A modified chequerboard titration demonstrated that LxA4 was predominantly chemokinetic whereas LTB4, as a positive control, was predominantly chemotactic.

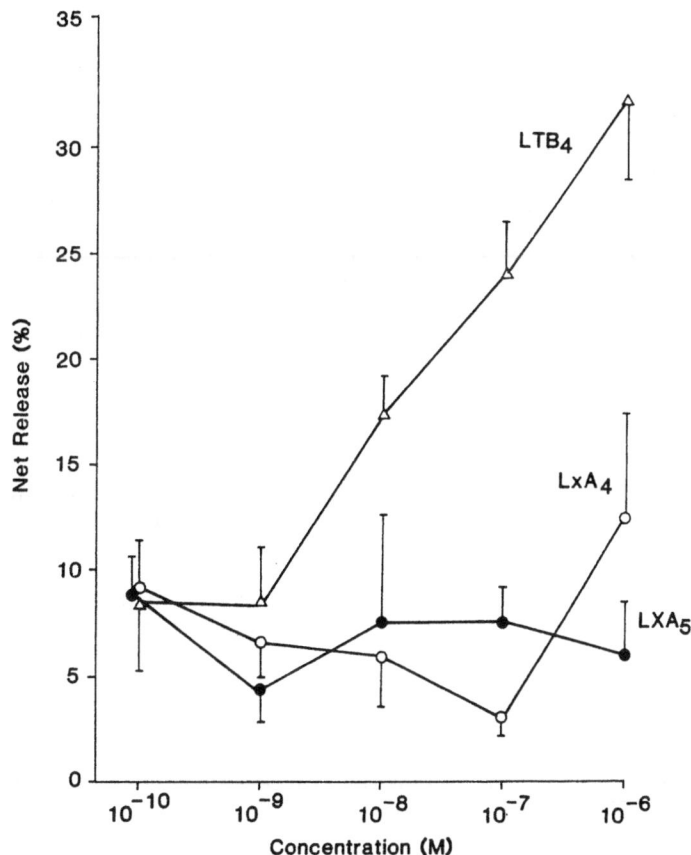

Fig.8 The Effects of LTB4, LxA4 and LxA5 on Lysozyme Release

Lysosomal enzyme release

The effects of LTB4, LxA4 and LxA5 were evaluated on lysozyme release. There was a dose dependent increase in lysozyme release from neutrophils induced by LTB4 from 10 M to 10 M. LxA4 and LxA5 were both unable to elicit neutrophil lysozyme release (see Fig 8).

CONCLUSIONS

We have prepared lipoxins derived from arachidonic acid and EPA by total organic synthesis using a novel synthetic route. We have demonstrated that LxA4 and LxA5 have intrinsic spasmogenic activities which were selective for guinea pig lung parenchymal strips over the concentration range studied. The activities of these novel molecules were slow in onset and long lasting. Lipoxins were substantially less potent than LTD4 and appeared to exert their effects on airways smooth muscle via an FPL 55712 inhibitable mechanism. LxA4 but not LxA5 induced chemokinesis of human neutrophils and the response appeared to have a high degree of stereoselectivity. LxA4 did not induce lysozyme release from human neutrophils in these experiments.

REFERENCES

1. P. Y. K. Wong, R. Hughes, and B. Lam, Lipoxene: A new group of trihydroxy pentaenes of eicosapentaenoic acid derived from porcine leukocytes. Biochem Biophys Res Commun. 126: 763 (1985).

2. T. H. Lee, R. L. Hoover, J. D. Williams et al., Dietary enrichment with eicosapentaenoic and docosahexaenoic acids in human subjects impairs in vitro neutrophil and monocyte function and leukotriene generation. N Eng J Med 312: 217 (1985).

3. B. Spur, A. Crea, and W. Peters, Synthesis of leukotrienes C5, D5 and E5, Arch Pharm (Weinheim) 317: 280 (1984)

4. J. Adams, B. J. Fitzsimmons, Y. Girard, Y. Leblanc, J. F. Evans, and J. Rokach, Enantiospecific and stereospecific synthesis of lipoxin A. Stereochemical assignment of the natural lipoxin A and its possible biosynthesis, J Am Chem Soc 107: 464 (1985).

5. J. Adams, B. J. Fitzsimmons, and J. Rokach, Synthesis of lipoxins: total synthesis of conjugated trihydroxy eicosatetraenoic acids. Tetrahedron Lett 42: 4713 (1984).

6. T. H. Lee, A. Crea, W. Peters, and B. Spur, Stereocontrolled total synthesis of lipoxin A5 and lipoxin B5 (submitted)

7. J. Morris and D. G. Wishka, Synthesis of lipoxin B, Tetrahedron Lett 27: 803 (1986).

8. K. C. Nicolaou, C. A. Veale, S. E. Webber and H. Katerinopoulos, Stereocontrolled total synthesis of lipoxins A. J Am Chem Soc 107: 7515 (1985).

9. K. C. Nicolaou and S. E. Webber, Stereocontrolled total synthesis of lipoxins B. Synthesis 453 (1986).

10. E. J. Corey, M. M. Mehrotra, and W-G Su, On the synthesis and structure of lipoxin B. Tetrahedron Lett 26: 1919 (1985).

11. B. Spur, H. Jendralla, A. Crea et al. Syntheses of leukotriene analogs Arch. Pharm (Weinheim) 319: 140-143 (1986).

12. B. Spur et al. Synthesis of deuterated lipoxins (submitted).

CONTRIBUTORS

Lutz Alder
Department of Chemistry
Humboldt University
1040 Berlin
Hessische Street 1-2
German Democratic Republic

Kamal F. Badr
Renal Division, Dept. of Medicine
Vanderbilt University
Nashville, TN 37232, U.S.A

Thure Björck
Department of Physiology
Karolinska Institutet & the National
Institute of Environmental Medicine
Stockholm, Sweden

Alan R. Brash
Division of Clinical Pharmacology
Vandrebilt University
Nashville, TN 37232, U.S.A.

Robert Brasseur
Macromolecules at Interfaces
CP 206/2
Brussels Free University
B-1050 Brussels
Belgium

Attilio E. Crea
Department of Medicine
UMDS, Guy's Hospital
London SEL, U.K.

Sven E. Dahlen
Department of Physiology
Karolinska Institute
Stockholm, Sweden

Michel Deleers
Research Center
UCB Pharmaceutical Sector
B-1420 Braine l'Alleud
Belgium

Lilian Franzén
Department of Physiology
Karolinska Institutet & the National
Institute of Enviromnental Medicine
Stockholm, Sweden

Brian J. Fitzsimmons
Merck Frosst Canada Inc.
Point Claire-Dorval, Quebec
Canada H9R 4P8

Jesper Haeggström
Dept. of Physiological
  Chemistry
Karolinska Institutet
Stockholm, Sweden

Crawford Jacques
Department of Medicine
UMDS, Guy's Hospital
London SEL, U.K.

Hartmut Kühn
Insitute of Biochemistry
Dept. of Pharmacology
Humboldt University
1040 Berlin, Hessische St.3-4
German Democratic Republic

Bing K. Lam
Dept. of Pharmacology
New York Medical College
Valhalla, NY 10595, U.S.A.

Tak H. Lee
Department of Medicine
UMDS, Guy's Hospital
London SEL, U.K.

Hisao Matsuda
Department of Physiology
Karolinska Institutet & the
Natl. Insti. of Environmental
  Medicine
Stockholm, Sweden

K.C. Nicolaou
Department of Chemistry
Univ. of Pennsylvania
Philadelphia, PA 19104
U.S.A.

John A. Oates
Department of Pharmacology
Vanderbilt University
School of Medicine
Nashville, TN 37232
U.S.A.

Tapio Puustinen
Department of Physiological Chemistry
Karolinska Institute
Stockholm, Sweden

Johan Raud
Department of Physiology
Karolinska Institute & the National
Institute of Environmental Medicine
Stockholm, Sweden

Joshua Rokach
Merck Frosst Canada, Inc.
Pointe Claire-Dorval, Quebec
Canada H9R 4P8

Bengt Samuelsson
Department of Physiological Chemistry
Karolinska Institute
Stockholm
Sweden

Charles N. Serhan
Hematology Division
Brigham and Women's Hosptial &
Harvard Medical School
Boston, MA 02115, U.S.A.

Bernd W. Spur
Department of Rheumatology and
Immunology, Brigham and Women's
Hospital, Harvard Medical School
Boston, MA 02115, U.S.A.

Natsuo Ueda
Department of Biochemistry
Tokushima University
School of Medicine
Tokushima 770
Japan

C.A. Veale
Department of Chemistry
University of Pennsylvania
Philadelphia, PA 19104
U.S.A.

S.E. Webber
Department of Chemistry
University of Pennsylvania
Philadelphia, PA 19104, U.S.A.

Pär Westlund
Dept. of Physiological Chemistry
Karolinska Institutet
Stockholm, Sweden

Rainer Wiesner
Institute of Biochemistry
Humboldt University
1040 Berlin
Hessische Street 3-4
German Democratic Republic

Eva Wikström
Department of Physiology
Karolinska Institutet & the Natl.
Institute of Environmental Med.
Stockholm, Sweden

Patrick Y-K. Wong
Department of Pharmacology
New York Medical College
Valhalla, New York 10595
U.S.A.

Shozo Yamamoto
Department of Biochemistry
Tokushima University
School of Medicine
Tokushima 770
Japan

Chieko Yokoyama
Department of Biochemistry
Tokushima University
School of Medicine
Tokushima 770
Japan

INDEX

injury, 108, 115, 152
microcirculation, 134, 135
micropuncture, 131
perfusion, 133
Glycosidation, 83
granulocyte (also see
neutrophils, eoinophils,
basophils)
isolation, 138
migration, 140, 153
responses, 138-144
Gross structure
of lipoxin A$_4$, 60
of lipoxin B$_4$, 60
Guinea pig
lung strip, 108, 115, 152
ileum, 118, 108

Hamster, 107, 108, 126
Hamster cheek pouch model, 10
Hematocrit, 132
HF-pyridine
High Pressure Liquid Chromatography
(HPLC) 17, 19, 23, 29, 41,
32-33, 55-57
HPLC profiles
of products form porcine
leukocytes, 31
products from 5-LO, 23
products from 12-LO, 19
products from epoxide
hydrolase, 9
lipoxin A$_4$, 9
trihydroxytetraenes from
reticulocyte
enzyme, 40-41
porcine leukocyte products, 53
separation of isomers, 87
Histamine, 115, 117, 119, 121, 123
Homeostasis, 58
Human
bronchi, 123
epoxide hydrolase, 7
lung, 124
Hydrogenation
selective, 62, 77
Lindlar, 69
cyclic carbonates in lipoxin
synthesis, 85
5-0-silyl-intermediates in
LXB$_4$, 87
Hydrolysis, 7
in glycoside intermediate, 83
total synthesis of lipoxins,
73, 85
enzymatic conjugate, 82
Hydrophilic centers
in lipoxins, 95, 100
Hydrophobic centers
in lipoxins, 95, 100

Hydroperoxy fatty acids
5-hydroperoxyeicosatetraenoic
acid (5-HPETE), 80
13S-hydroperoxy-9,11(Z,E)-
octadecadienoic acid, 40
15-hydroperoxyeicosapetaenoic
acid (15-HPEPE), 28, 52
15-hydroperoxyeicosatetraenoic
acid (15-HPETE), 3, 39, 80,
88
methyl ester, 39-45

Ileum, 10, 108, 122
muscle, 111, 118
Indomethacin, 115, 152
Inflammation, 12
Inflammatory cellular infiltrates,
131
Inhibitors
lipoxygenase, 39
nafazatrom, 118
nordihydroguaiaretic acid (NDGA),
118
Injury, 30, 131
Interactions
LXA$_4$ and LTC$_4$, 120
Interfacial, 97, 101-103
Intermediates
aldehyde, synthesis of, 64
biological, 7
in biosynthesis
epoxide, 58, 91
5(6) epoxide tetraene, 7
leukocytes, 3
mechanism, 39
model for, 9
Intermediates continued
in total synthesis of
LXA$_4$, 62, 64, 65, 83, 85,
89-91
LXB$_4$, 68, 69, 82, 85, 87-91
Intermolecular S$_N$2, 63
Intracellular role, 1
Intravital microscopy, 10
Inulin, 132, 133
Ionophore (A23287), 3, 52, 100, 137
Isolated vascular strips, 110
Isolation of
Lipoxins of the four series, 1-3
Lipoxins of the five series,
147-150
Isomerization, 4
Isomers
Lipoxin A$_4$, 3, 4, 55, 62, 85,
139-142
6(S)-lipoxin A$_4$, 4, 85, 141
11-trans Lipoxin A$_4$, 4, 66,
113, 114, 140
Lipoxin B$_4$, 3, 4, 33, 67, 80,
139-140